Global Economic Cc

The New Dark Ages:

The Glass Banking Pyramid

Perhaps the key question to ask in this crisis is: Do we have an actual financially based market failure (supply and demand) or is it leadership based government failure (rules and regulations)?
~Neal Russell Vanderstelt

Table of Contents

Introduction

You will not get a left right bias out of me. I present my opinion right down the middle, criticizing stupid moves by government, and look at it sensibly, analyzing causes and looking for practical solutions. However, the problem has escalated and the solutions will probably fall out of what is realistically possible at this stage of the game.

The opportunity to learn from the past and apply it to the future is here. Would you rather lose all your money and be unprepared for whatever blind fate may drop in your path? When you see banks closing it is too late to prepare. Don't wait until there are bank runs and panics before taking action, when banks go out of business, or war is imminent, and you lose everything. It will be too late to position yourself financially and prepare for any disruptions in society. The key to knowing the future is in knowing the past and learning what clues to look for.

In 2008, most banks went into crisis mode, after the collapse of the housing market. It was a much greater problem than just a housing market crisis; it was a massive financial crisis, most were not prepared in the slightest. Society had repeated the same mistakes of the past and had not learned the critical lesson. Even people who did not lose all of their wealth, lost a considerable portion.

The economic statistics across the world are fraudulent data. It is hard to know what is going on by listening to mainstream news. The conditions are much worse and can be seen in the increasing poverty levels and geopolitical instability. These mistakes are constantly repeated from our past. It is a pattern of human nature that is repeated by ignorant leaders who fail to seek knowledge, or just don't care about the fate of society. Since leaders are often ignorant and greedy; we must learn beforehand and prepare for the worst.

The title reflects what I think will happen in the coming years and the unstable nature of money, credit, and debt, which are all politically connected to incompetent leaders who have sold us out. The left and right paradigms are a conjured up fake, reality disguising what really goes on behind the making of the global ponzi-schemes. Some of the stuff I talk about may not sound like anything new if you watch alternative media or read in between the lines. The dots are connected, without resorting to conspiracy theories that could be true, but have yet to be proven.

Some basic terms should be understood such as GDP, debt-to-GDP, the differences between the national debt and unfunded liabilities, QE, bank runs, the US Debt Ceiling, free market economy, a fiat currency, and austerity. A basic understanding is good enough. Other terms can be looked up. The following partial definitions were taken from Wikipedia under each individual term. To get the full definition search each individual term (e.g. "GDP" or "QE").

Gross domestic product (GDP) is a measure of the size of an economy. It is defined as "an aggregate measure of production equal to the sum of the gross values added of all resident, institutional units engaged in production (plus any taxes, and minus any subsidies, on products not included in the value of their outputs)"

In economics, the debt-to-GDP ratio is the ratio between a country's government debt and its gross domestic product (GDP). A low debt-to-GDP ratio indicates an economy that produces and sells goods and services sufficient to pay back debts without incurring further debt. Geopolitical and economic considerations - including interest rates, war, recessions, and other variables - influence the borrowing practices of a nation and the choice to incur further debt.

Government debt (also known as public debt, national debt and sovereign debt) is the debt owed by a central government.

According to the U.S. Debt Clock, total long term Unfunded Liabilities are at $126 trillion, a $1.1 million liability for each U.S. taxpayer. The main driver of that astronomical number is two of our major entitlement programs: Social Security and Medicare. Unfunded liabilities are an expense that does not have savings or investments set aside to pay it.

Quantitative easing (QE) is a type of monetary policy used by central banks to stimulate the economy when standard monetary policy has become ineffective. A central bank implements quantitative easing by buying specified amounts of financial assets from commercial banks and other financial institutions, thus raising the prices of those

financial assets and lowering their yield, while simultaneously increasing the money supply. This differs from the more usual policy of buying or selling short-term government bonds in order to keep interbank interest rates at a specified target value.

A bank run (a run on the bank) occurs when, in a fractional-reserve banking system (where banks normally only keep a small proportion of their assets as cash), a large number of customers withdraw cash from deposit accounts from a financial institution at the same time because they believe that the financial institution is, or might become, insolvent; and keep the cash or transfer into other assets, such as government bonds, precious metals or stones. When they transfer funds to another institution it may be characterized as a capital flight. As a bank run progresses, it generates its own momentum: as more people withdraw cash, the likelihood of default increases, triggers further withdrawals. This can destabilize the bank to the point where it runs out of cash and thus faces sudden bankruptcy. To combat a bank run, a bank may limit how much cash each customer may withdraw, suspend withdrawals altogether or promptly acquire more cash from other banks or from the central bank, besides other measures.

The United States debt ceiling limit is a legislative limit on the amount of national debt that can be issued by the US Treasury, thus limiting how much money the federal government may borrow. The debt ceiling is an aggregate figure applies to the gross debt, which includes debt in the hands of the public and in intra-government accounts. (About 0.5% of debt is not covered by the ceiling.) Because expenditures are authorized by separate legislation, the debt ceiling does not directly limit government deficits. In effect,

it can only restrain the Treasury from paying for expenditures and other financial obligations after the limit has been reached, but which have already been approved (in the budget) and appropriated.

In a free market economy, prices for goods and services are set freely by the forces of supply and demand and are allowed to reach their point of equilibrium without intervention by government policy. It typically entails support for highly competitive markets and private ownership of productive enterprises.

Fiat money is currency which derives its value from government regulation or law. It is not backed by a physical commodity.

Austerity is a set of policies with the aim of reducing government budget deficits. Austerity policies may include spending cuts, tax increases, or a mixture of both.

Geopolitics focuses on political power in relation to geographic space. In particular, territorial waters and land territory in correlation with diplomatic history. Academically, geopolitics analyses history and social science with reference to geography in relation to politics.

The next chapter will begin with modern times and how we ended up with such an out of control global paradigm. We will start at the early roots and work our way to recent times to fully understand the situation. You'll find the history very interesting. *Statistics used are official government data unless otherwise stated.

Chapter 1 The Roaring 20's

I. Alcoholic beverages were prohibited from 1920 - 1933.

II. Ford assembly line inspires mass production.

Back in the Roaring 20's, the economy was good, and nobody cared about the future of the economy. Everyone thought there could be no end. It was an unrealistic attitude and many bought into it with every penny they had. Life was ruff back then, however, invention improved conditions, and everyone wanted a taste of the new creations. They were willing to sacrifice all to get into the game.

What really started the roaring 20's? It just didn't happen by luck. We need to understand this to understand what works and what doesn't in our society. Various factors contributed to the roaring 20's: World War I had finally ended, people were eager, and had new desires. Prohibition (1920-1933) kicked off the 1920's and there were new social networks created from it. Black jazz music clubs were roaring in Harlem and people came from all over to hear the festive shows. There was not only an American Dream but blacks too had their version of the Black American Dream. It was a very happy time for most. American women won the right to vote in 1920 and there were new hopes. The enthusiasm was off the scale.

III A vintage Ford Truck

The earlier invention of the Automobile jump started the 20's economy of the United States, thanks to Henry Ford, and his assembly line factories. It reduced the cost of a car to $360 while Ford paid his workers $5 a day (in 1914). It was good pay back then and workers could also afford to buy a Ford. 72 days at $5 a day = $360 (enough to buy a Ford). This concept was referred to as Fordism. There were strings attached to the $5 a day pay. Some of the pay came in bonuses, rather than a check, but it still allowed many working class people a chance to buy a car. The car helped most families, businesses, and different parts of society. The first car (a little buggy) was invented in 1908 and by 1929 half of all Americans owned a car.

The culture and every aspect of society changed since many now had a car. People (especially young) could go out on dates to the movies in cars, everyone's morale improved, people shopped more, and the economy had a fire brewing. People were now spending more money as a result and had their eyes on everything. As consumerism began to take shape in the United States, due to modern inventions, culture itself changed, and became more technology based. The invention of the automobile, the movie picture, and the radio created major industries that thrived in tandem. A mass culture was created out of the radio and movies. The entertainment industry began to take off as did the news stations.
New corporations began to form and sports legends were created by this mass culture.

Many corporations that were created in the 20's are still thriving today. Broadcast networks, NBC, and CBS were created in 1926. Although, Ford Corporation started in 1908, Henry Fords manufacturing concepts influenced other businesses and improved production of the 20's throughout various industries. The United States became the richest country in the world.

Credit or installment buying became the way of the American consumer. It was a new concept. They could take items home without paying for them and make payments over time. Advertising and credit seemed to go hand and hand. Buy now, pay later. By 1927, 75% of all households good were acquired through credit. It created overproduction because people could buy more than they could afford.

Stocks in corporations began to soar. Many people profited from a booming stock market. Everyone had an interest in owning stocks. People believed investing in stocks was respectable and reliable. Market manipulation took place and insiders (in particular) made fortunes. They would pump up a particular stock, by buying it themselves and recommend the stock, then dump. Smaller speculators began to gamble in the markets, risking everything. They were buying on margin in large numbers -- creating an irrational frenzy to buy stocks.

You could buy a $100 stock for $25 on margin and possibly make $200 or more. This influx caused the stock market to soar. Stocks alway seemed to go up, up, and up. There was no belief the market could crash and people continued to buy, buy, buy.. Technology made this all possible. From 1921 to 1929, the Dow Jones rocketed from 60 to 400, creating many new millionaires.

Everything was pretty much unregulated in the 20's and there was a lot of fraud. People perhaps began having too much fun. Not producing enough to fulfill their demands and enjoyments. It was a great time but then came 1929. Paul Warberg (a Wall Street Banker) broke ranks and warned of a depression in March 1929. His warning fell on deaf ears. Warberg died only a few years later after issuing a warning to Americans.

IV. Paul Warberg

Chapter 2

1929 Stock Market Crash

V. Dismayed crowd gathers at Wall Street after crash. Just imagine everyone losing everything they own at the same time.

On September 3, 1929, the Dow Jones Industrial Average reached a record high of 381.2. US stock prices began to fall on September 4, 1929.
The complete collapse didn't happen in a single day. There were waves of selling spread out through the days of September and October in 29'. It persisted to be a bear market for many years. Speculators and investors are continually reassured everything is okay. October 24th, 1929 (black Thursday): by lunchtime, RCA is down 35%, and Montgomery Ward has nose-dived 39.9%. Investors are reassured everything is okay.

October 28th selling picked up again and in 36 hours of time, after the open on October 29th, stocks and American industry itself were now worth 22% less in only 2 days of trade. All the people who bought on margin had to put up collateral to the brokers because of losses. Waves of selling continued. On October 29th, 1929, panic selling escalated; it was called Black Tuesday.

Overproduction of goods without having the money to pay for them weakened the real worth of the markets. The crash was completed in 1932; stocks had lost nearly 90 percent of their value. The stock market fell to its lowest point during the depression on July 8, 1932. The irrational exuberance of the Roaring 20s was over.

October 30, 1929, John D. Rockefeller, Sr. announces: "There is nothing in the business situation to warrant the destruction of values that has taken place on the exchanges during the past week. My son and I have for some days been purchasing sound common stocks." The Dow had one of its best days ever, rocketing up 29 points. The Dow would go on to lose 84.1% more of its value by July 8, 1932.

People hysterically rushed to the banks to get their savings, fearing the banks would run out of money. Their worst fear came true; the banks were out of money.
In only 1 year 800 banks went under. All over the country people lost not only their money, they also lost their jobs.

Dow Jones Industrial Average Big Losses 1929-1932			
Date	Change (ticks)	% Change	Closing Price
Oct. 28, 1929	-38.33	-12.82%	260.64
Oct. 29, 1929	-30.57	-11.73%	230.07
Nov. 6, 1929	-25.55	-9.92%	232.13
Sept. 24, 1931	-8.20	-7.07%	107.79
Aug. 1932	-5.79	-8.40%	63.11

Quotes During and After The Crash:

(The following quotes are from Wikiquotes – to see more go to the website)

1929 "Wave after wave of selling again moved down prices on the Stock Exchange today and billions of dollars were clipped from values". "Traders surged about brokerage offices watching their holdings wiped out.... It was one of the worst breaks in history.".... "For a time, in the morning, the market was showing signs of rallying power.... Then new waves of selling out of poorly margined accounts started another reaction..." Minneapolis Star account of "Black Tuesday" (October 29, 1929), when the stock market "collapsed": panic selling took place, with owners of stock wanting to sell, no matter how great their loss.

Sept, 1929 "There is no cause to worry. The high tide of prosperity will continue." Andrew W. Mellon, Secretary of the Treasury

Dec 5, 1929 "The Governments business is in sound condition." Andrew W. Mellon, Secretary of the Treasury

January 13, 1930 "Reports to the Department of Commerce indicate that business is in a satisfactory condition, Secretary Lamont said today."

Jan. 24, 1930 "Trade recovery now complete President told. Business survey conference reports industry has progressed by own power. No Stimulants Needed! Progress in all lines by the early spring forecast." New York Herald Tribune

March 8, 1930 "President Hoover predicted today that the worst effect of the crash upon unemployment will have been passed during the next sixty days". Washington Dispatch

June 29, 1930 "The worst is over without a doubt." James J. Davis Secretary of Labor.

Chapter 3

The Great Depression In America

VI. Unemployment percentage

before and after The Great Depression.

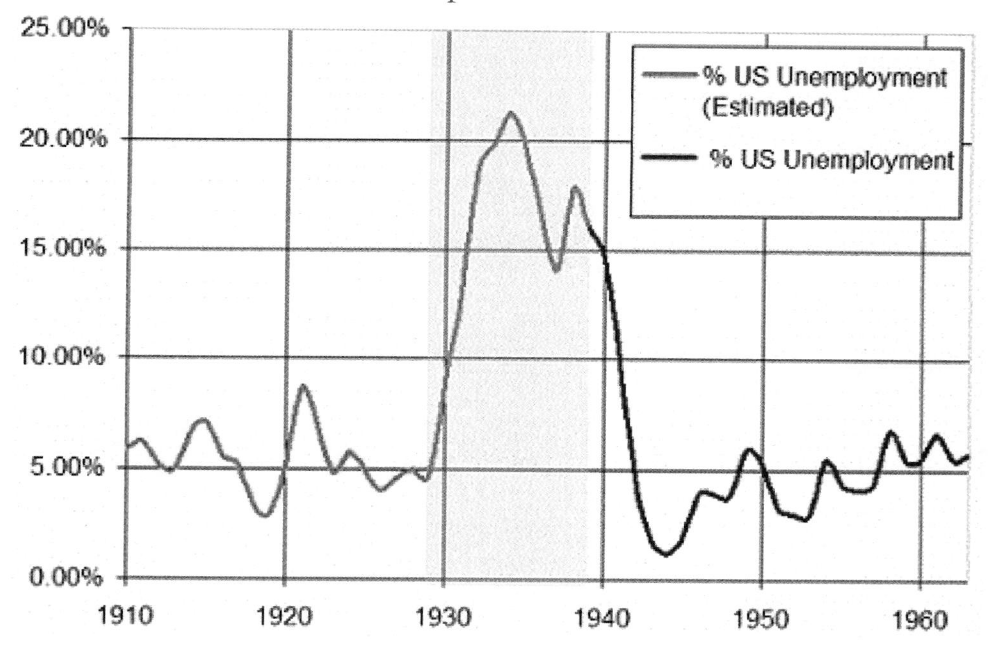

After the stock market crash unemployment in the United States rose to 25% (at peak), wages fell 42%, economic growth fell 50%, and world trade plummeted 65%. By 1931, 200 banks had closed.

Many people took to soup lines to eat. Men were dressed poorly and some had no shoes so they wrapped things on their feet or if they had shoes they were usually worn out and they would have to reinforce them. Work was often paid in barter

such as a pair of shoes or wood to keep warm. Clothes and other commodities were all hard to come by and a good percentage of the population suffered. A great number of people lost their homes and had to live in unfit conditions. Some lived in makeshift shelters made of cardboard and other scavenged materials. People in more deprived areas would even fight over garbage.

Farmers were swimming in debt and many foreclosed. Between 1930 and 1931 farm prices fell by more than 30%. Price of wheat, cattle, milk, and hogs, had all collapsed and caused great stress for farmers. Farms were also hit by an eight year drought in the 30's.

Bank runs due to panic are what actually started The Great Depression. No more cash and no more loans meant the golden age of the 20's was completely gone. It was a time in history never seen and it was pretty much unregulated. Many banks as a result closed their doors.

Year	Commercial Bank Closures	Deposits $
1929	659	230,643
1930	1,350	837,096
1931	2,293	1,690,232
1932	1,453	706,187
1933	4,000*	3,596,708

Day after day people would look for work and get declined. Some would work, after being promised pay, and not paid after doing the work. Many were paid through barter.

My grandfather told me the food trucks were looted and they had to have a guard ride the trucks with baseball bats to knock off looters from produce trucks. He did a lot of work thinking he was going to get paid, and at the end of a full day of work, the guy gave him a log of wood. He had a car from when times were good, but could not afford to put gas in it, so he rode a bicycle to look for work. Every day he would look but there was no work and he would end up in a soup line.

Chapter 4 Hooverville

VII. Pictures of makeshift housing called Hooverville's.

A "Hooverville" was a shanty town built by homeless people during the Great Depression. They were named after Herbert Hoover, who was President of the United States during the onset of the Depression and was widely blamed for it.

Military Vets were denied bonus pay from World War I (they were given a rain check to be paid in 1945), so they protested for the pay they earned. Many were out of work and homeless, due to the Depression. They set up makeshift living structures out of scavenged materials and garbage, near their protest areas.

Most of the Bonus Army camped in a Hooverville on the Anacostia Flats, across the Anacostia River from the federal core of Washington. The marchers remained at their campsite waiting for President Hoover to act. On July 28, 1932, Attorney General William D. Mitchell ordered the police to remove the Bonus Army veterans from their camp. When the veterans moved back into it, police drew their revolvers and shot at the veterans, William Hushka and Eric Carlson were killed.

Hoover and General Macarthur declared war on the protesting veterans. Macarthur sent in military, including tanks against the veteran protesters. They cheered as the saw the military intervention, believing it was a parade in honor of them. After the cavalry charged, the infantry, with fixed bayonets and tear gas entered the camps, evicting veterans, families, and camp followers. The veterans fled across the Anacostia River to their largest camp, and President Hoover ordered the assault stopped.

MacArthur chose to ignore the president and ordered a new attack, claiming that the Bonus March was an attempt to overthrow the U.S. government. Fifty-five veterans were injured and 135 arrested. A veteran's wife miscarried when 12-week-old Bernard Myers died in the hospital, after being caught in the tear gas attack. They savagely burned their shelters down. If you ever questioned whether the military would attack civilians, you now understand they not only attacked civilians, but they even attacked military vets. Military people typically have families and hearts, but in the end they are trained to simply follow orders.

Hoover would not help individuals, he would only help businesses. He was a hardcore capitalist, for the corporations. It was up to charities to help the needy, was his belief, and it "undermined" the American way. Poverty and unemployment were best left to "voluntary organization and community service." The state helping individuals was a "European philosophy," Hoover contended.

The Bonus Army Stamp

The Bonus Army At the U.S. Capitol

Though the Bonus Army incident did not derail the careers of the military officers involved, it proved politically disastrous for Hoover. He lost the 1932 election in a landslide to Franklin D. Roosevelt.

Chapter 5

The Great Depression Worldwide

The depression was led by the United States but it affected the world. Worldwide GDP (Gross Domestic Product) fell by 15% from 1929 to 1932. Germany was one of the hardest hit countries. After losing World War I, they were now facing a dire economic situation along with war debts. It would spur World War II. If one had $1000 on September 3rd 1929, it would have gone down to $108 by July 8th, 1932.

Economic Changes 1929–32	United States	Great Britain	France	Germany
Industrial production	-46%	-23%	-24%	-41%
Wholesale prices	-32%	-33%	-34%	-29%
Foreign trade	-70%	-60%	-54%	-61%
Unemployment (increased by)	+607%	+129%	+214 %	+232%

The Great Depression in Germany

Germans didn't rely on American production, they were an industrial country, but they relied on American loans. The country was already in bad shape before the depression, hyperinflation had occurred due to war reparations, and the occupation of the Ruhr (its main industrial center) further crippled the country. The economic condition turned dire. No

further loans were issued to Germany in late 1929, while American financiers began to call in existing loans. As a result of the recalled loans the German economy collapsed. As banks collapsed, the savings of many Germans were wiped out. Extreme views of the Communist and Nazis suddenly became more popular.

(As seen in chart) German Marks depreciate rapidly in Weimar Republic. It took only 3 yrs to impoverish most Germans.

The hyperinflation in the Weimar Republic was a three-year period of hyperinflation in the Weimar Republic (modern-day Germany) between June 1921 and January 1924. It took a barrel of money to buy a loaf of bread and some Germans burned piles of money as newspaper for fires. There was famine conditions and extreme poverty. These subjects are typically only lightly touched on by English based historical literature.

The Great Depression in Italy

The Great Depression hit Italy very hard, despite its agricultural base. International trade declined, as did hours worked, production, investment, and wages. Italy had emerged from World War I in a poor, weakened condition, and under fascism from 1922-1945. As industries came close to failure they were bought out by the banks, largely by illusionary bail-outs, and the assets used to fund the purchases were largely worthless. The Italian financial crisis peaked in 1932.

The Great Deprssion in Greece

Prior to the Great Depression, the Greek economy experienced years of growth.. When the Depression struck, banks and local businesses faced unpayable loans and plummeting asset values. By 1932 central bank reserves had dropped so much that they only backed 40% of Greek bonds. Like Germany and Italy, Greece was devestated.

The Great Depression in Australia

Australia was hit by the decline in demand for wool and food exports. Throughout the 1920s the Australian unemployment rate floated between 6% and 11%. By 1932 almost 30 percent of Australian workers were without a job. Like most countries, Australia suffered years of high unemployment, poverty, low profits, deflation, and plunging incomes.

(As seen in picture) There were incidents of civil unrest, particularly in Sydney, Australia. In 1931, over 1000 unemployed men marched from the Esplanade to the Treasury Building in Perth, Western Australia to see Premier Sir James Mitchell.

The Great Depression in France

The Great Depression affected France from about 1931 through the remainder of the decade. The crisis hit France in 1931, a bit

later than other countries. "The winter I spent on the streets - the winter of '32-33 - was no milder nor harder than any other winter; the winter cold is like labour pains - whether it lasts for a longer or shorter period of time there is always the same amount of pain. That particular winter, it snowed and it froze; thousands of young men, forced out of their jobs by the crisis, struggled on to their last penny, to the end of their tether then, in despair, abandoned the fight...
On street benches and at métro entrances, groups of exhausted and starving young men would be trying not to die. I don't know how many never came round. I can only say what I saw. In the rue Madame one day I saw a child drop a sweet which someone trod on, then the man behind bent down and picked it up, wiped it and ate it." -An unknown Frenchman

The depression in France was not accompanied by a banking crisis, only one major bank failed. There was no major banking crisis in France. In 1931, many countries decided to devaluate their currency. The French government rejected the option of devaluation and capital controls.

France, traditionally a right-wing government, decided to elect a left-wing government in 1936. It was called the Popular Front led by Léon Blum. It included radical members, and French Communist Party members. When the Radicals would not accept currency controls, unrest, and capital flight occurred. It weakened the economy and employers tried to minimize the results of the Matignon agreement, which created more social tension and in turn a further flight of capital. The Franc (French Currency) was devalued by 30%. However, the economy continued to stall. By 1938 production still had not recovered to 1929 levels, and higher wages had been neutralized by inflation. Businessmen took their funds overseas. Blum was forced to stop his reforms and devalue the franc. With the French Senate controlled by conservatives, Blum and thus the whole Popular Front, fell out of power in June 1937.

* The Matignon Agreements of 1936, an agreement between the French government, employers and labor guaranteeing trade union membership and negotiating rights, a 40-hour working week and paid workers' holidays. At the same time the French currency was also devalued at this time.

The Great Depression in Great Britain

Britain had never experienced the boom that had characterized the U.S., Germany, Canada and Australia in the 1920s, so their market bust was less severe. Britain's exports fell by half (1929–33), the output of heavy industry fell by a third, and unemployment increased severely in most sectors. Hardest hit by economic problems were the industrial and mining areas in northern parts of England. Unemployment affected most workers although it was not extended to agricultural workers until 1936.

The Great Depression in Japan

Japan was not severely effected by The Great Depression. The Japanese economy shrank by 8% during 1929–31. By 1933, Japan was already out of the depression which credited success was ironically do to a currency devaluation. However, Japan took steps to immediately cool off the economy to slow inflation after the recovery.

The Great Depression in Russia (The Soviet Union)

The Soviet Union was the only communist state with very little international trade. The economy was not tied to the rest of the world and was only slightly affected by the Great Depression.

Chapter 6

Dollar Devaluation

In The Great Depression

During the 30's, most countries abandoned the gold standard, resulting in currencies that no longer had intrinsic value. With widespread high unemployment, devaluations became common. Effectively, nations were competing to export unemployment, a policy that has frequently been described as "beggar thy neighbor." The US Dollar was depreciated by 41% in 1934. Gold was basically stolen from citizens and given to the government in 1933, and then the gold was given to the US Treasury. [It is talked about in more detail in a later chapter.] The dollar ultimately was devalued by 69% through gold manipulation during The Great Depression.

England was considered the reserve currency of the time, and (like the US) devalued its currency by 30% in 1931, and again by 40% in 1934. England was the first country to devalue and it put pressure on other countries to follow. France then followed with a devaluation of the French Franc by 30%, the Italian Lira was devalued by 41%, and the Swiss Franc by 30%. Most European countries devalued and Greece defaulted in 1932.

Large currency depreciations instantly hit consumer purchasing power and reduce wages. In the case of The Great Depression, it resulted in plummeting national incomes, shrinking demand, mass unemployment, and an overall decline in world trade.

The "LOGICAL BELIEF" of this system is that the government could spend its way out of economic hardships with borrowed money. "I've got a lot of debt so I'll just make more"! One look at the historical purchasing power of the US dollar for example, will tell you a different story since the US dollar is worth 99% less today. It would be like you borrowing 1000 bucks from the person you owe, to pay 1000 bucks of accumulated interest, and continually doing it over a 100 year period and while the interest goes up, borrow more to pay just the interest. This is pure insanity and it seems to be orchestrated to keep people poor and the rich richer. Remember, it is not their money they are using, it belongs to society.

Currency Devaluation Per Country Great Depression	Devalued by
United States	69%
England	40%
France	30%
Italy	41%
Switzerland	30%
Australia	40%

The gold standard was in place to keep central banks from printing too much money. If they did people could simply turn in their cash for gold.

Chapter 7

Gold Is Made Illegal

5,000 years ago in Egypt and the Middle East: gold and other metals were used as money. **In 225 BC, the Roman Empire used the first gold coins. In 1774 the British Parliament introduced the gold standard.**

UNDER EXECUTIVE ORDER OF THE PRESIDENT

Issued April 5, 1933

all persons are required to deliver

ON OR BEFORE MAY 1, 1933

all GOLD COIN, GOLD BULLION, AND GOLD CERTIFICATES now owned by them to a Federal Reserve Bank, branch or agency, or to any member bank of the Federal Reserve System.

The Bank of England decided to take the British Pound off the gold standard in September 1931. The US would follow suit under President Roosevelt (aka FDR). One begins to wonder about how bad Hoover really was in comparison to what would happen next with FDR's stunts. Maybe not as bad as 1st perceived?

Until 1933, people carried gold coins in their pockets, and paper bills were exchangeable for gold and silver coins at any bank. After the system popped, gold was made illegal. Many people are not even aware of this, but yes indeed, the government took everyone's gold, and gave it to the Federal Reserve Bank. On January 30th, 1934, the Gold Reserve Act required the Federal Reserve to give its gold and gold certificates to the US Treasury. This caused a 41% devaluation of the US dollar. The government essentially increased the value of its gold by 69%. For a time, gold reserves actually exceeded the money supply, and 75% of gold was owned by the central banks by 1948.

The penalty for not turning in your gold was a $10,000 fine and/or 10 years in prison. Americans were legally ROBBED of 41% of their wealth. Perhaps the greatest heist of individual wealth in history took place and it was all government backed.

With Executive Orders the government has dictatorial powers. And you think they won't be abused? An EXECUTIVE ORDER is not voted on, it does not go through other channels, 1 person decides.

Chapter 8

World Wars and War Debts

Hitler was on the rise and England wanted to get him. Germany had great success in rebuilding under the furor. Adolf Hitler put many Germans back to work. It was such a turn around it was wrote about as a great success throughout the world. We can't have that now can we?

Not to praise Hitler, but what he did in shame generally came after WWII was declared, and even the US was hesitant to get involved. Each country committed its fair share of atrocities, which could have possibly been avoided, and there should have been every attempt to avoid such a catastrophe. Millions died and many are paying the debts to this day. Just think, if one sole assassin could have taken out the furor, instead of lining up every army in the world, as if pieces on a chess board, and creating massive global war debts and costing the death to millions.

Wars come at great cost, because a bomb blowing up always has a negative effect on the economy. A bomb explodes, everything is immediately lost, and it is 100% debt creation. The cost of war is 100% debt. Even things gained in war, such as technology; generally don't outweigh the cost of war, especially modern wars with modern weapons.

The US never recovered after the World Wars. In fact, the US never paid off the debt going back to the civil war. Before the World Wars, the debts were manageable. Just how far has this new technology really got us? We are obviously not sustainable. Will we evolve fast enough to keep up with gains in technology before destroying humanity as we know it or will the New Dark Ages be upon us?

Chapter 9

The Bretton Woods System

The IMF (International Monetary Fund) and World Bank are created.

With World Wars, The Great Depression, massive global debts, and massive devaluations of currencies, there had to be a monetary change. All Allied countries met in the Bretton Woods Conference in 1944 before the end of World War II. The US was made the global reserve currency, because it was the largest economy, and represented 50% of the world's economy. The United States, which controlled two thirds of the world's gold, insisted that the Bretton Woods system rest on both gold and the US dollar.

The Bretton Woods system of monetary management established the rules for commercial and financial relations among the world's major industrial states in the mid-20th century. The Bretton Woods system was the first example of a fully negotiated monetary order intended to govern monetary relations among independent nation-states. The chief features of the Bretton Woods system were an obligation for each country to adopt a monetary policy that maintained the exchange rate by tying its currency to gold and the ability of the IMF to bridge temporary imbalances of payments.

The Bretton Woods system ushered in a period of prosperity and rapid economic recovery. It did not last because The Federal Reserve Bank (FED) continued to print money and there wasn't enough gold to back the newly created dollars.

Prior to the new system the British pound was considered the reserve currency because other central banks held British bonds as reserve assets. The US is now officially the new king on the block. For a time, Bretton Woods worked out, but watch what people with power do.

Chapter 10

The Nixon Era Of Fraud

The Watergate and Vietnam era and yet it gets even uglier.

Republican Richard Nixon was found guilty of the Watergate scandal and he escalated the Vietnam War. In the 1968 election, Nixon claimed to have a plan to end the war in Vietnam, but, in fact, it took him five years to disengage the United States from Vietnam. Nixon is also attributed to creating the drug war (by declaring a war on drugs) and creating the DEA. The prison population has grown out of hand since that time and the US leads the world in prison population.

Gold no longer backs the US dollar as of August 15th, 1971. This deregulated the US dollar -- it was the cause of massive future inflation. At the time, gold was valued at approximately $35 an ounce, and now such a low price for gold cannot even be imagined. Since the gold standard was removed, the buying power of the US Dollar has gone done substantially.

But what was the real reason behind this? It was a technical default by the United States because of massive war debts from the world wars and the new wars, Korea and Vietnam. The US continued to create debt for the Vietnam War.

The French under Charles De Gaulle wanted to be paid back and turned in their cash for gold because the French didn't believe the US could honor debt obligations. It proved to be true. Nixon declared a Force Majeure, a form of default, and removed the gold standard. Cash could no longer be converted to gold. Prior to, you could turn in your cash for gold, and it helped regulate the currency from excessive debt creation.

Since 1933 gold was illegal in the US and typically only governments owned large amounts of it. After the decoupling of gold, the average person is now forced to borrow, and go deeply into debt. It caused by the creation of a fiat currency, a currency not backed by anything. An unlimited amount of dollars and debt could now be created. Nixon had allowed for a currency that would essentially be fictional in nature, with no backing. The trend of fiat money and massive debt creation is still ongoing to this day.

Chapter 11

Gold and Deregulation of Money

Deregulation began in the 70's as stated earlier with the Nixon Shock. Decoupling the dollar from gold isn't the only thing that deregulated paper money. There are interest rates in play and government legislations. Below, is a FED rate historical chart:

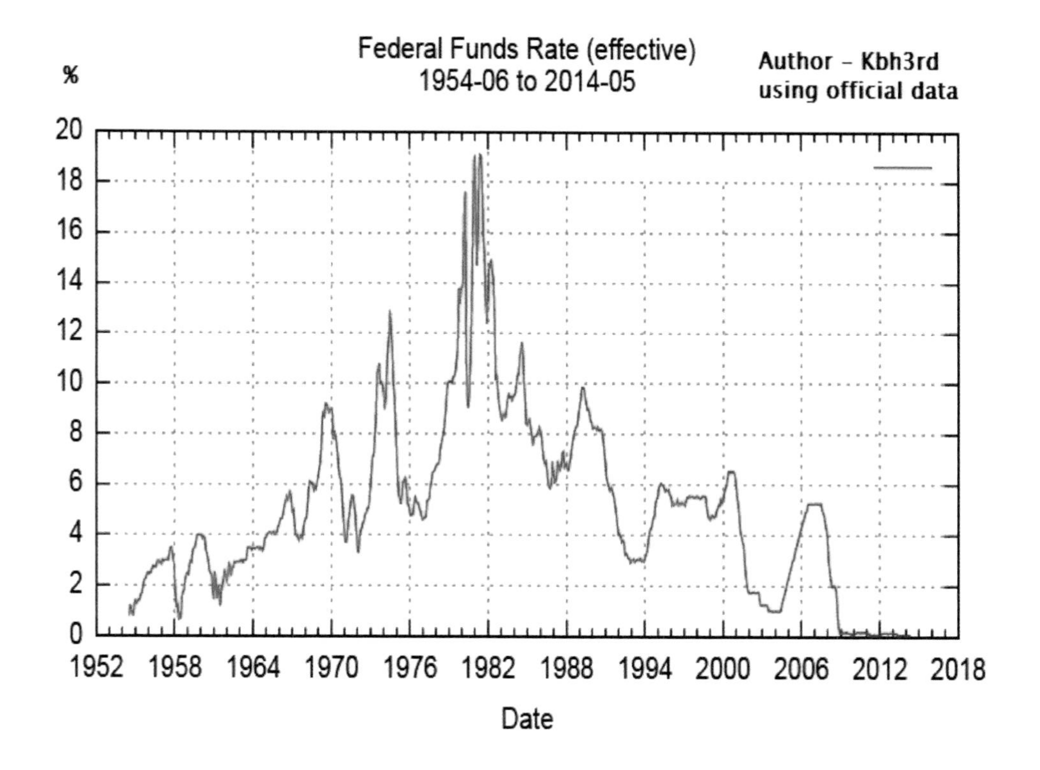

As the FED bank lowers rates it tends to depreciate the value of money (to supposedly stimulate the economy). Raising the rates cuts inflation down, while lowering it causes inflation. After 2008, the rates have been essentially dead. In a healthy economy, the rates should not be too high or too low.

After the flat rates of 2008, QE was also implemented. The Federal Reserve creates dollars and injects them into the economy? That's what they say at least but when has the FED showed up at your door and given you some cash after an injection? The money is not distributed equally----other people received your portion.

Nobody actually knows where this money goes. Since the monetary base (amount of money in circulation) is increased your value of money goes down.

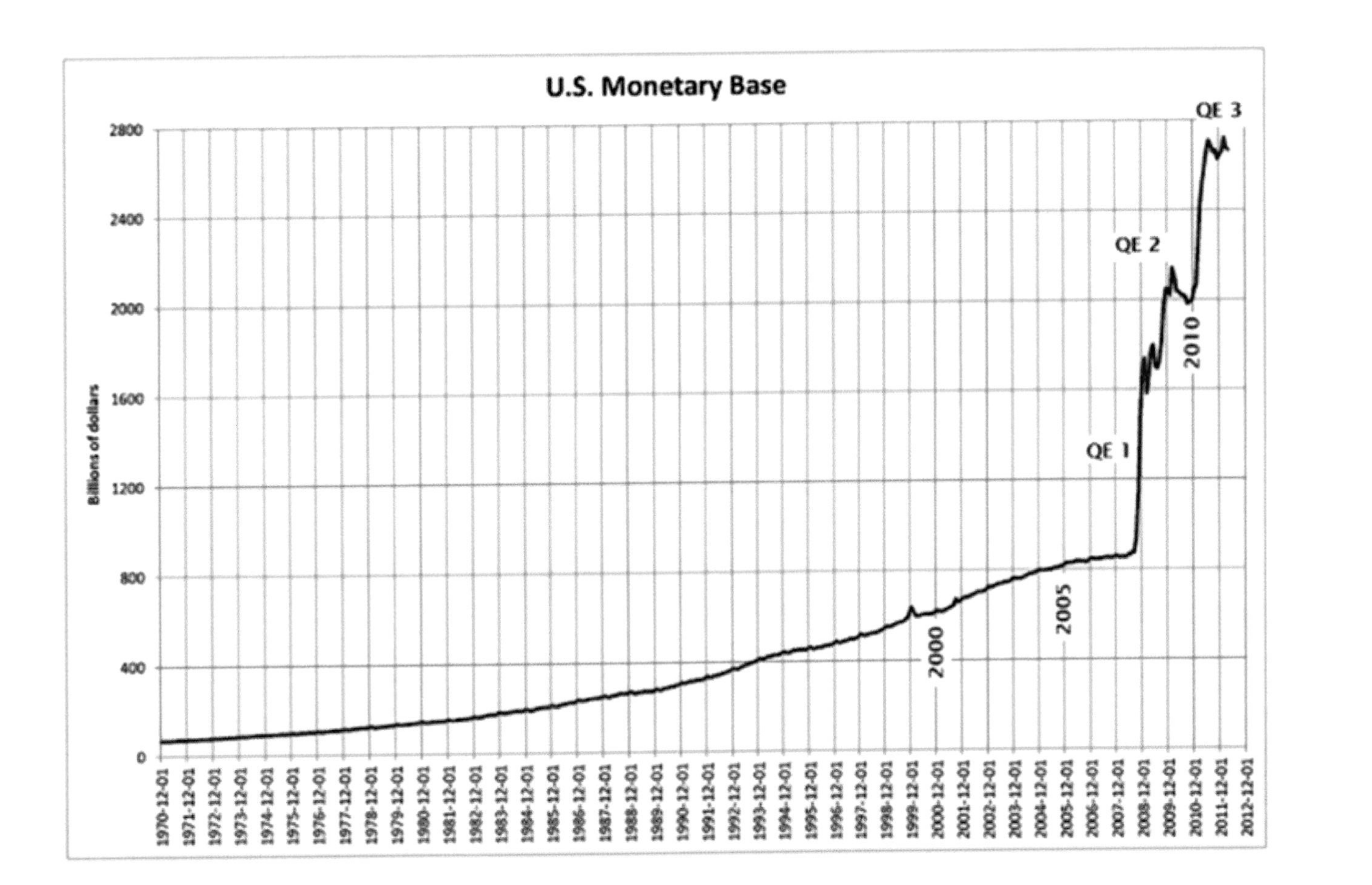
U.S. Monetary Base
Billions of dollars
0
400
800
1200
1600
2000
2400
2800
QE 1
QE 2
QE 3
2000
2005
2010
1970-12-01
1971-12-01
1972-12-01
1973-12-01
1974-12-01
1975-12-01
1976-12-01
1977-12-01
1978-12-01
1979-12-01
1980-12-01
1981-12-01
1982-12-01
1983-12-01
1984-12-01
1985-12-01
1986-12-01
1987-12-01
1988-12-01
1989-12-01
1990-12-01
1991-12-01
1992-12-01
1993-12-01
1994-12-01
1995-12-01
1996-12-01
1997-12-01
1998-12-01
1999-12-01
2000-12-01
2001-12-01
2002-12-01
2003-12-01
2004-12-01
2005-12-01
2006-12-01
2007-12-01
2008-12-01
2009-12-01
2010-12-01
2011-12-01
2012-12-01

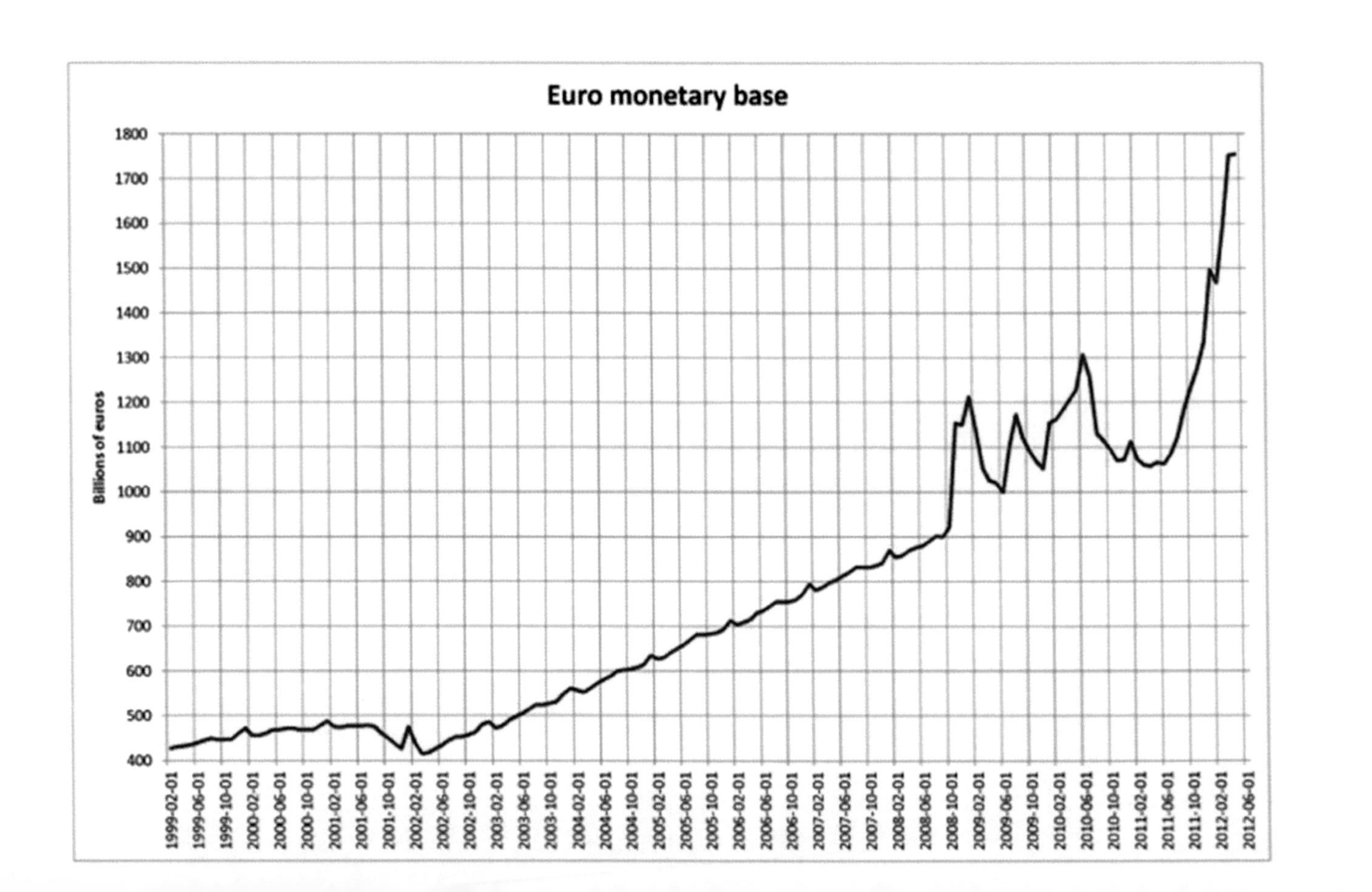

Euro monetary base
Billions of euros
1800
1700
1600
1500
1400
1300
1200
1100
1000
900
800
700
600
500
400
1999-02-01
1999-06-01
1999-10-01
2000-02-01
2000-06-01
2000-10-01
2001-02-01
2001-06-01
2001-10-01
2002-02-01
2002-06-01
2002-10-01
2003-02-01
2003-06-01
2003-10-01
2004-02-01
2004-06-01
2004-10-01
2005-02-01
2005-06-01
2005-10-01
2006-02-01
2006-06-01
2006-10-01
2007-02-01
2007-06-01
2007-10-01
2008-02-01
2008-06-01
2008-10-01
2009-02-01
2009-06-01
2009-10-01
2010-02-01
2010-06-01
2010-10-01
2011-02-01
2011-06-01
2011-10-01
2012-02-01
2012-06-01

Gold had been banned in 1933, after the gold standard was removed in 1971, it was then legalized in 1974. Sounds like manipulation to me! Why couldn't you own gold when the dollar was backed by gold? Well of course that gold was only for the government. Now time to inflate it perhaps? They never gave it back to the families that had their gold repossessed. Theft is legal when the government does it.

After gold was legalized the price went up substantially, because there was no gold standard – the dollar was not backed by anything other than the paper itself, and whatever the government chose to do. The amazing thing was that the government owned most of the gold. The fact of the matter was that the price increases were due to dollar devaluation, because the dollar was a free floating currency, and essentially unregulated. It was not an increase in value due to supply and demand, but rather more and more dollars were in circulation, while debts continued to accumulate. Unemployment and inflation went up through the 70's and into the 80's, while low economic growth was the norm. They called it "stagflation".

In the 60's and 70's, the US no longer enjoyed uncontested economic, political, and military dominance over the world. Politically, it started to become more unstable, and there was the Vietnam War, among other conflicts. This war trend became the new norm.

The question is: do you have a right to a job, or should you have to earn that right, by being a hard or skilled worker? Obviously, disabled people need help from society, so I'm not asking the question to be cruel, but able bodied and mentally efficient people are a different story.

Missing from the puzzle that many people ignore is The Employment Act of 1946. It set up the continual growth of cash and credit in the name of promoting employment and declaring employment as a right. It all sounds good doesn't it? The problem with all this is the currency was still unregulated so the government can just make changes on the fly. The more meddling by the government, the more long term poverty is created by kicking the can.

Unemployment in the 70's began to rise -- a new act was created that explicitly instructs the nation to strive toward four ultimate goals: full employment, growth in production, price stability, and balance of trade and budget. So the government did more meddling and created Resolution 133, allowing for more manipulation of money to "stimulate" the economy. Reference: Federal Reserve Reform Act of 1977.

.

Chapter 12

Banks and Housing

Get Fully Deregulated

The Glass Steagall Act distinguished investment banks from regular deposit banks. In 1999 it was repealed under Bill Clinton, by the Gramm-Leach-Bliley Act. Without the Gold Standard there wasn't much regulating the money system other than Glass Steagall. When it was repealed it gave regular banks the green light to gamble with customer deposits, as if you deposited your money into a risky investment fund under 10-1 leverage. So for every dollar you deposited the banks could gamble 10 on leverage. Under Clinton The Commodity Futures Modernization Act exempted credit-default swaps from the regulation.

On signing The Gramm-Leach-Bliley Act, President Clinton said that it, "establishes the principles that, as we expand the powers of banks, we will expand the reach of the [Community Reinvestment] Act." Bill Clinton wasn't done setting up a future catastrophe for the economy with deregulating. He passed a bill that would allow basically anyone to qualify for a home loan by rewriting The Community Reinvestment Act. It put added pressure on banks to lend to low-income neighborhoods. It set up the housing bubble -- many people foreclosed on their homes -- causing massive poverty

After the housing crash of 2008, many banks got rewarded, rather than punished due to being "too big to fail." It would be all fine and well, if the corporations who are fully deregulated, faced punishment through a true capitalist system. In other words, they go out of business, and allow for new competition; that wasn't the case. Many Republicans supported the deregulations, as well as bailouts. President George W. Bush signed into law the Emergency Economic Stabilization Act of 2008, which of course included the bailouts of financial corporations, banks, and later automakers.

Troubled Asset Relief Program (TARP) under Bush, was the catalyst for Obama's stimulus plan. TARP allowed the Treasury to purchase illiquid, difficult-to-value assets from banks and other financial institutions. The targeted assets can be collateralized debt obligations, which were sold in a booming market until 2007, when they were hit by widespread foreclosures on the underlying loans. A tarp is used to cover up junk -- I guess they had some humor there (i.e., cover up junk assets). Unfortunately, these junk assets cost tax payers $40 billion of new debt. The government spent roughly $475 billion in TARP bailouts, most of which has been supposedly paid back. So the $40 billion that is being written off amounts to a loss of less than 10 percent on the original expenditures.

Banks were now allowed to run wild, and in terms of the home loans, were forced to give loans to people who did not really qualify. History is in the making for a future catastrophe.

Chapter 13

Destruction of Infrastructure

Many products are not made in the US and are outsourced. The value of the dollar is a reflection of a country's labor. The US went from producing product to being a consumer based nation. A country as large as the US cannot be a consumer nation and expect to survive. Especially with the amount of war engagements -- something will have to give. Up until the 70's the USA also produced oil, and even exported it. The US now relies on foreign oil.

China invested in its infrastructure, while the US let the infrastructure erode. We are talking roads, bridges, buildings, and factories. In engineering terms, many roads and bridges do not get a passing grade in the US. Many jobs are outsourced. This not only wrecks the ability of a country to produce things, but also how desirable it is for rich foreigners that typically help expand nations. The quality of living in the US is in a nosedive.

The US is basically a dead industrial giant that used to be the king of the world for workforce and production. Birthrates have declined and countries like China and India have a population that are 4x bigger (8x combined) than the US. That's a lot of labor force and people in these countries will do anything for work. Americans cannot compete with their labor forces. As if it wasn't enough, these countries have deregulated their labor laws and have very loose environmental laws.

Chapter 14

The 2008 Crisis Is A Depression

In our fiat fake money system, most of the money in our economy is created by banks when they make loans. But, in the aftermath of the financial crisis, banks stopped lending, and so stopped creating new money.

QE1, QE2, QE3, need I say any more? QE (Quantative Easing) only postpones the inevitable crash that will occur in the future. The purpose of QE was to supposedly stimulate the economy so people and businesses could get loans. But it had very little effect, because many people did not qualify for these loans, since the economy was bad, and banks are very reluctant to lend. QE only served as a fancy manipulation of the bond market and prolongs a future crash. This manipulation has a price because they are putting in good money and making bad debt. The future crash will now be amplified by more toxic debt.

Despite the trillions of dollars already expended recapitalizing banks, there is very little, if any, progress to show. Instead there is just more toxic waste and junk bonds, while very little goes into the real economy.

Real Unemployment is over 20%. Google it! It has been since 2008 and this fact has not changed. For a brief period in The Great Depression, unemployment spiked to 25%, but did not stay above 20% for as long as the current crisis. Real Unemployment (as calculated before Bill Clinton changed how it was calculated), has stayed above 20% since 2008, and that's depressonary! It is now late in 2015 as this is being written. Bank rates are flat, there has been a massive amount of injections in the form of QE and other manipulations, there is a record number of food stamp recipients, many people are still being laid off, and not much has changed. The housing market still struggles and people have a very hard time making the ends meet. The food stamp situation is equivalent to the amount of people in soup and bread lines during the 30's. There is a massive amount of people living on the streets. The media has ignored the situation. It is a dire situation and the poverty rate is magnifying.

11 trillion dollars was spent in QE by the world's central banks since the 2008 crisis. It could have paid off most of the U.S. household debt – all mortgages, all student loans, all car loans, and credit cards. In that time, GDP has decreased. The "positive" side is US stocks have been in a bull run while many people around the world are impoverished. Stocks as of this writing are the most overvalued in history.

With 11 trillion we could probably have a colony on mars, while you may laugh, with the coming wars we will probably need another planet to live on. The QE is a signal that everything has been tried and failed. This is the desperate hail marry attempt to try to fix things. Well it is not working either.

The government is buying the debt by printing dollars to buy it, while interest rates are at zero. This will not only create more debt but it will set up the future scenario where the government has to raise the rates to curtail inflation. That will create a higher premium on the interest on debt. Creating more debt, creates more quicksand.

Chapter 15

Funny Math

Bad Economic Statistics

Chart for Alternative Unemployment is -- Courtesy of ShadowStats.com

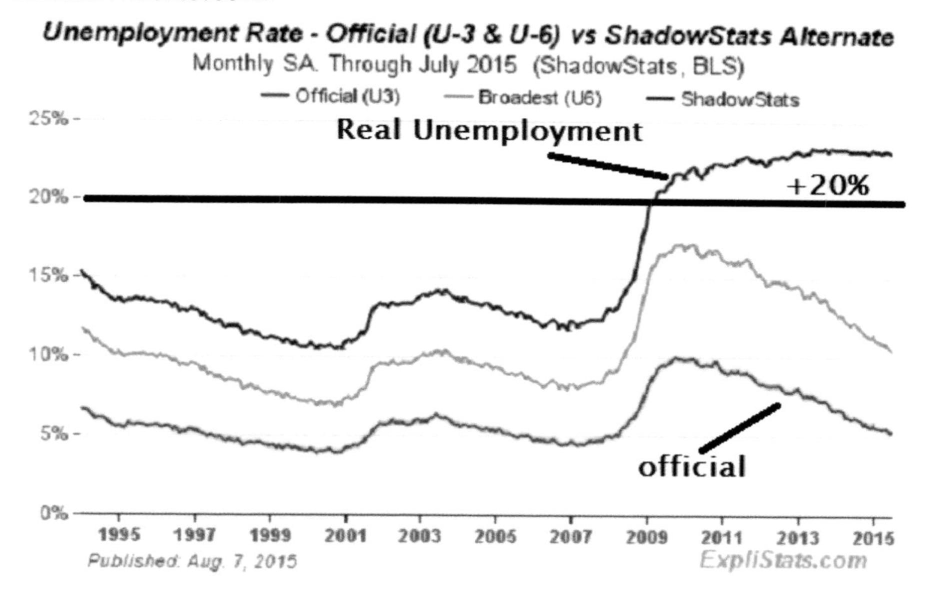

The true unemployment is much higher than the official numbers. The reason being is under Bill Clinton, the way unemployment is calculated was changed. Using the old way, unemployment is as much as 4x higher. Real unemployment is thought to be over 20% (google: alternative unemployment or real unemployment). Could it be worse than the great depression and we don't even feel it because of technology? We don't see the bread or soup lines because there are EBT credit cards. There are a record number of food stamp recipients. That signals a depressionary scenario and proof enough that there was no recovery after 2008.

Labor statistics (Labor Participation) confirm the unemployment figures do not make sense. If unemployment improved, so would the amount of labor, but that isn't the case.

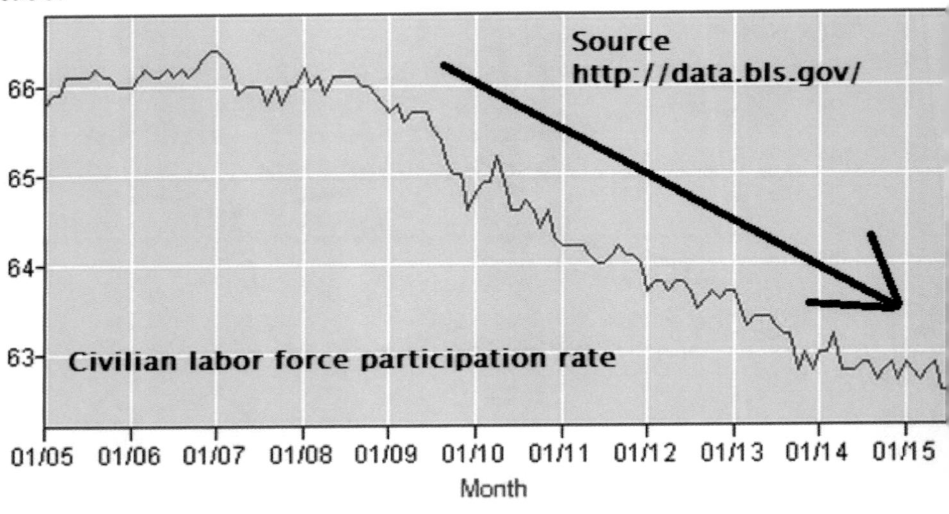

Meanwhile, the government continues to spend, make debts, and expand government. The problem is, it doesn't have this money.

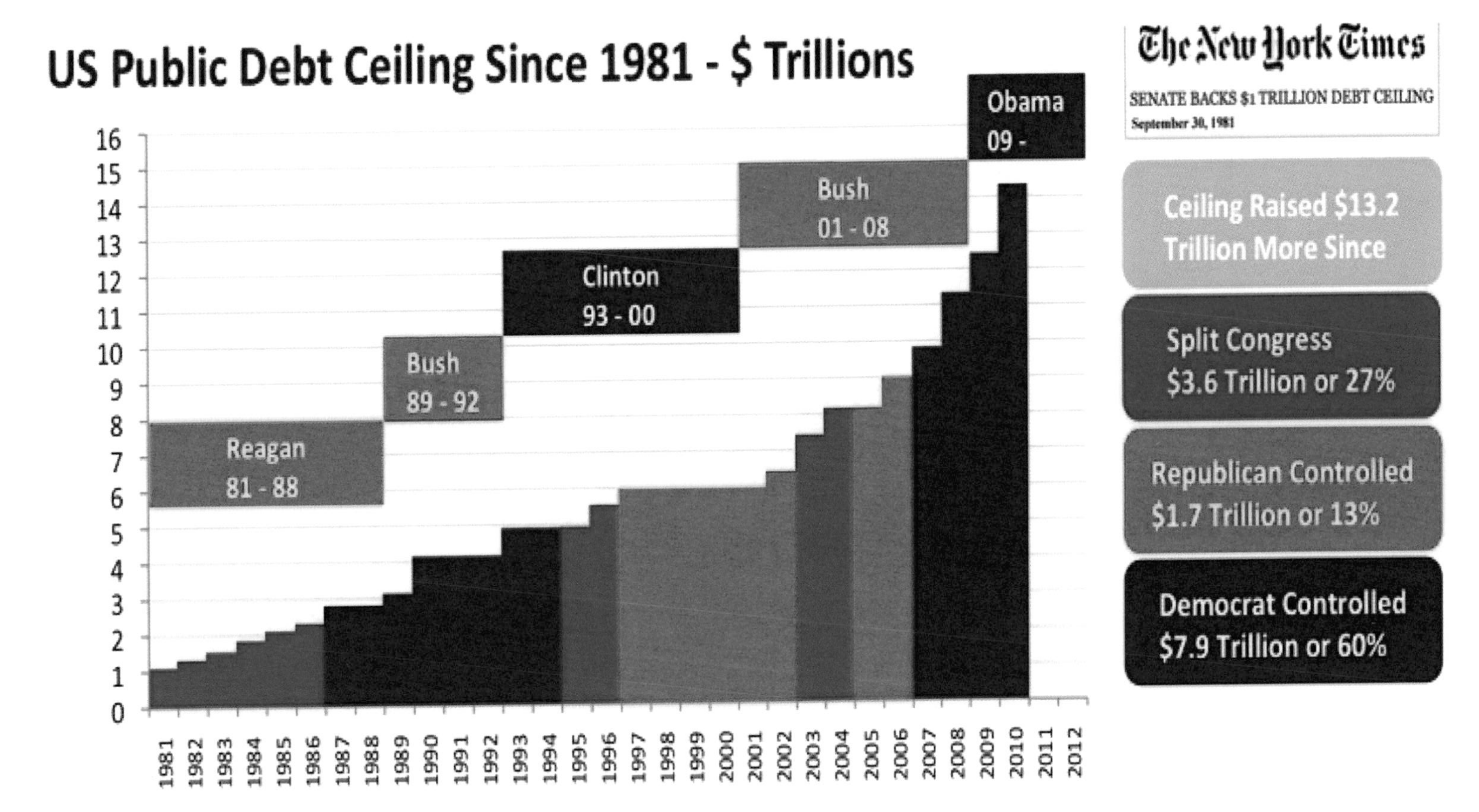
US Public Debt Ceiling Since 1981 - $ Trillions
16
15
14
13
12
11
10
9
8
7
6
5
4
3
2
1
0
Reagan
81 - 88
Bush
89 - 92
Clinton
93 - 00
Bush
01 - 08
Obama
09 -
1981
1982
1983
1984
1985
1986
1987
1988
1989
1990
1991
1992
1993
1994
1995
1996
1997
1998
1999
2000
2001
2002
2003
2004
2005
2006
2007
2008
2009
2010
2011
2012
The New York Times
SENATE BACKS $1 TRILLION DEBT CEILING
September 30, 1981
Ceiling Raised $13.2
Trillion More Since
Split Congress
$3.6 Trillion or 27%
Republican Controlled
$1.7 Trillion or 13%
Democrat Controlled
$7.9 Trillion or 60%

The debt ceiling is not funded, it is a debt. The government creates artificial debt to pay itself because technically the US is bankrupt. The entire reason there is a debt ceiling is that fact. If the US had the money it would not need an imaginary credit card. The increase of the debt ceiling also reflects the erosion of the USD in action. Trillions seem to mean very little to monetary policy makers. Just imagine if you had your own credit card, and you could increase your amount every time you exceeded the limit – that is exactly what the government routinely does and it devalues the buying power of our money.

How nice are these new banking rules? When you deposit money in the bank you are considered an unsecured creditor to the bank and it is basically their money. As an unsecured creditor you stand last in the queue of creditors to be paid out of any funds and or assets which the bank has to pay its creditors. This is part of a deal made with the G20. Many people are unaware of it.

By this time the can has been kicked so far down the road, it cannot be found. I think it might be in space somewhere floating about with the rest of the multi-trillions of debt and quadrillions in derivatives!

Chapter 16

World War III

The Iran Deal, Iran-Israel tensions, and dislike of the US.
Iran goes nuclear and has severe dislike of both the US and Israel. Along with other well known disputes in the Middle East. This conflict takes a priority because of the nuclear threats Iran has made. Iranian leaders have many times expressed their willingness to nuke Israel off the map. There have been many anti-Israel and anti-American protest held in the streets of Iran. While many claim the possession of nuclear energy does not lead to nuclear weapons – one look at the world and you can see that most nations who have nuclear technology have nuclear weapons.

Iran has also invested in long-range missile technology making it a greater threat. The Iranian nuclear deal (to say the least), has made the Israelis angry, and they are willing to go in preemptively. The scenario does not look good -- no matter how you look at it. In my opinion, it is just a matter of time, before something breaks out that will involve Iran (either directly or indirectly), and there are already many disputes going on in the Middle East that could drag the Iran-Israel conflict onto the table. All hell will break out should the latter be the case. The Middle East is oil region and the entire globe will be involved should there be multiple full scale wars. That is WW-III scenario #1.

Israel-Syria conflicts

These are ongoing conflicts and wars over an area called the Golan Heights, an area captured by Israel in 1967. According to the Bible, an Amorite Kingdom in Bashan was conquered by Israelites during the reign of King Og. The western two-thirds of this region are currently occupied by Israel. Since 1964, Arab countries, concerned over Israeli plans to divert waters of the Jordan River into the coastal plain, had been trying to divert the headwaters to deprive Israel of water resources, provoking tensions between Israel on the one hand, and Syria and Lebanon on the other. The conflicts are known as the 1967 Six Day War and 1973 Yom Kippur War. In a Six-Day War, Israel defeated Jordan and captured the West Bank, defeated Egypt and captured the Gaza Strip and Sinai Peninsula, and defeated Syria and captured the Golan Heights.

On October 6th, 1973, as Jews were observing Yom Kippur, Egypt and Syria launched a surprise attack against Israeli forces in the Sinai Peninsula and Golan Heights, which opened the Yom Kippur War. The war ended on October 26th, with Israel successfully repelling Egyptian and Syrian forces. Conflicts in the region are ongoing and can lead to much bigger disputes, especially considering that Israel is the only predominant Jewish country in the region and are near many anti-Israel countries. Israel is thought to have nuclear weapons, but the details are unknown.

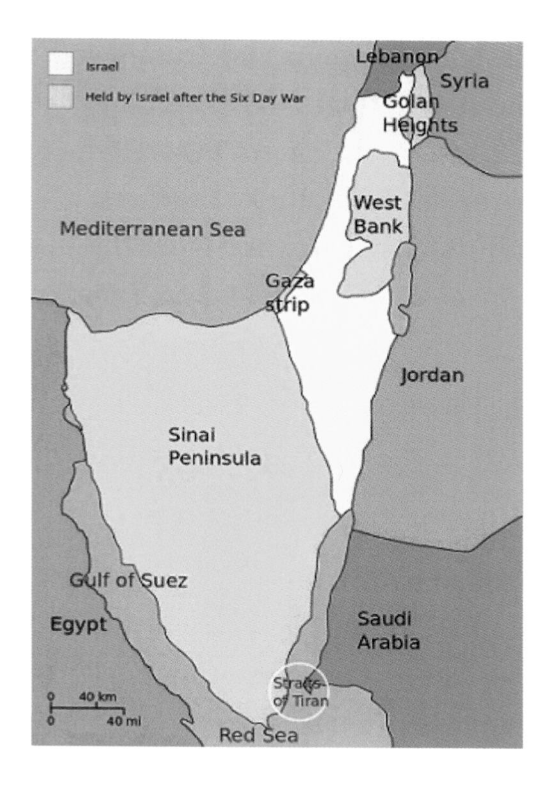

The Ukraine Crisis.

The breakup of the Soviet Union has always been a controversial topic. The closer you get to Russia, the more influence there is by Russia. Russia is the head Slavic country so to speak. It is their domain. Crimea (a land mass in Ukraine), was pretty much a Russian country before the crisis began, but western Ukraine was in dispute.

*More Russians died in World War II than any other people. They were the 1st going into Berlin, Germany to end the war by taking over Hitler's bunker.

The crisis stemmed over a political uprising in Ukraine, because the pro-Russia Ukrainian leader (Viktor Yanukovych) was ousted on February 21, 2014 through a series of ongoing protests. The protests had lasted many months. Russia sent in military intervention and claimed Crimea as Russian territory. The area was pretty much a Russian settlement before tensions arose, occupied by mostly Russian or pro-Russia settlers. There was already Russian military base there before the dispute. Russia was punished, Russian assets taken, sanctions applied, and the Ruble crashed.

A new wave of Anti-American sentiment spread stimulating Russian's to have extreme dislike of the western nations (primarily The US and Europe). Russia had also previously attacked Georgia in the Georgian conflict over disputes back in 2008. Many of the problems with Russia almost always relate back to Gazprom (a Russian energy corporation), and their energy sales. It is not always the case, however in any Russian-based dispute you have to read in-between the lines and find out if there is a deal that has to do with natural gas or oil.

Russia is a very aggressive country that will defend its business assets, especially in regions near Russia. While Russia is in economic trouble, the possibility of war is always greater, because people that are poor are more willing to fight. Russia has lots of poor Russians, who are very angry at America, and they are a militaristic society. Don't rule out a big one between the US and Russia. Even though the dispute has to do with Ukraine, with the American interference, it now is in-between Russia and the US. This conflict is ongoing and is a WW-III scenario.

Russia US tensions continue.

Putin in official speeches has stated that Russia is a nuclear power and would defend itself. On July 4th, Russia sent a nuclear capable bomber 50 miles off the coast of California. What was the message there? Russia unlike other countries is not only nuclear, but capable of hitting any country in the world. Russia was the 1st into space with satellites and very capable technologically.

Greece Crisis -- The ticking debt time bomb.

After World War II, Greece experienced an economic boom but that ended in the 1970's after political changes. In the 80's, unemployment began to climb, while inflation remained high. It went on into the 90's and government debts ballooned. The government tried to jumpstart the economy with deficit spending policies. It stayed a rocky road for Greece to this day.

Today, Greece's Debt-to-gdp is among one of the worst in the world. The flat rates of the Euro's financial policies do not account for the fact that Greece is not an industrial country. Greece is being held hostage to Germany under strict conditions to keep the Greeks from defaulting. The recent offering of a 3 yr deal by the European Union (EU) is to supposedly save Greece but it is under Draconian conditions. These conditions will harm future generations of Greeks.

The deal was rejected by 61% of Greeks, but they are being forced into it by corrupt leadership. They have protested in the streets, but their protest fall on death ears. Should the Greeks leave the Euro it could cause a domino effect. Italy, Spain, and Portugal, would likely all follow suit, due to hurting economies, and high unemployment. In my opinion, the policies of the EU are primarily intended to strongly favor Germany and France.

By joining the Euro, Greece was also able to borrow money at cheap rates (that Germany qualifies for), and that just ended up spelling disaster for Greece. It was like declaring happy hour for the local drunks. Prior to joining the Euro, Greece had economic problems that were not settled. It was a big mistake. The EU policies do not compensate for the geographic, cultural, and industrial differences between the countries. Greece is a tourist country, not an industrial giant like Germany, but it is Germany who holds their leash. There is instability in the Euro and it isn't confined to Greece. Many countries have similar problems as Greece, although Greece is probably the worst off, and they get put under the microscope.

Greece has defaulted before, but it was not part of the Euro, and has to follow the policies of the Euro. They also cannot adjust their fiscal policies to fit the nature of their own regional problems. It is just a matter of time before Greece explodes and this will likely cause a chain reaction, in the Eurozone from other suffering members.

European unemployment.

While overall it might not seem that bad, there is a huge problem with Youth Unemployment in the EuroZone. In some areas, youth unemployment is 50%, such as in Greece and Spain. This contributes to crime and it severely burdens the future chances for generations to come. The youth are the backbone of a country so this is a huge story. Young people are the big buyers in a society, while older people tend to buy less, so high youth unemployed has multiple negative effects. Experienced workers are more in demand.

Europe has many new immigrants, who contribute to unemployment and welfare services. When things get bad there will likely be a new wave of nationalism, which is already going on now throughout Europe.

Mexico border problems.

The US has relied on cheap labor for its big corporations, and through lobbyist, demanding for loose border restrictions. Mexican refugees have poured into the US with strong disapproval by many Americans. It has led to the extreme views by Presidential candidate Donald Trump to build a wall and deport all undocumented immigrants (also called illegal's). His extreme views have actually captured strong support. The problem isn't just the demographic change to America. Many feel that they have contributed to crime, are costing the nation due to an increased use of social services and taking of jobs that should be for legally registered Americans, who have to pay taxes. Supporters of open borders argue that the jobs they take are not jobs Americans will do and that the crime contribution is exaggerated. Mexico by contrast, has much stricter requirements, and policies for immigration.

In Mid-August of 2015, Mexico made a public announcement for Trump to build his own wall. Some Mexican extremist also believe that the US is Mexico's territory – especially the Southwest US that used to belong to Mexico, even though Mexico had sold it to the US. While Mexico is not a direct military threat it is a backdoor to terrorism and there could be future alliances with enemies of the US should Mexico become unstable. The Mexican economy is in big trouble, there is widespread poverty, and a large gap between the rich and poor.

China Japan disputes over assets and long steaming anger from WW2.

The Nanking Massacre or Nanjing Massacre, also known as the Rape of Nanking or Rape of Nanjing, was an episode during the Second Sino-Japanese War of mass murder and mass rape by Japanese troops against the residents of Nanjing (then spelled Nanking), then capital of the Republic of China. The massacre occurred over six weeks starting December 13, 1937, the day that the Japanese captured Nanjing. Since most Japanese military records on the killings were kept secret or destroyed shortly after the surrender of Japan in 1945, historians have not been able to accurately estimate the death toll of the massacre. The International Military Tribunal for the Far East estimated in 1948 that over 200,000 Chinese were killed in the incident. The Japanese had committed war crimes that have not been forgiven by the Chinese people. There is much tension especially since the Chinese are a traditional people who don't forget quickly. Many Chinese woman were raped by the Japanese and many were brutally tortured and killed. It is well found resentment. The Japanese are a much different society today, but there are other disputes which are financially based.

The Senkaku Islands dispute is one such dispute between China and Japan who both make claims to the territory. It is close to key shipping lanes, rich fishing grounds, and possible oil reserves. The island is also a strategic military location. In September 2012, the Japanese government purchased three of the disputed islands from their "private owner', prompting large-scale protests in China. On November 23, 2013, the PRC set up the "East China Sea Air Defense Identification Zone" which includes the Senkaku Islands, and announced that it would require all aircraft entering the zone to file a flight plan and submit radio frequency or transponder information. The islands are controlled by Japan. China says that the islands have been part of its territory since ancient times, serving as important fishing grounds administered by the province of Taiwan. Taiwan also claims the islands.

In November 2013, China also announced the creation of a new air-defense identification zone, which would require any aircraft in the zone – including the islands -- to comply with rules laid down by Beijing. China also has blamed the US for escalating tensions over the dispute. On July 30th, 2015, Col. Yang Yujun, a spokesman for the Ministry of National Defense, said it was the United States, not China, that was to blame for rising tensions in the resource-rich South China Sea.

China does quite a bit of currency manipulation of the US bond market and this could technically be seen as an economic attack against the US. It has repeatedly dumped US treasuries which has caused massive volatility in the financial markets. China is a major holder of US debt through the bond market. Without foreign purchasing of US bonds (also called treasuries) US debt and ultimately all US assets have no support. The fall of the US dollar would lead to Geopolitical crisis because many other currencies are pegged to the US dollar since the Bretton Woods agreement. China is trying to become the world's reserve currency, attempting to displace the USD, and has signed into agreements with other nations to do so. There is a definite Currency War going on -- that alone could lead to a WWIII like scenario -- should the dollar collapse and there is Geopolitical instability.

The New Development Bank (NDB), formerly referred to as the BRICS Development Bank, is a multilateral development bank operated by the BRICS states (Brazil, Russia, India, China and South Africa) as an alternative to the existing American and European-dominated World Bank and International Monetary Fund. The BRICS countries represent over 3 billion people, or 42% of the world population. This is substantial alliance against the USD, and in my opinion, is a strategic partnership to stage a collapse of the USD.

On the other hand, massive hedge funds also manipulate the price of gold, oil, and other commodities, which may be a financial attack against China and Russia, to destabilize the regions. Either way you look at it, a currency war is occurring, and we are all affected by it.

North Korea tensions with South Korea and the US for intervention.

Gojoseon (Joseon) was an ancient Korean kingdom and was invaded by the Han dynasty of China that led to its collapse into many states. There came into being, what is known as: The Three Kingdoms (Silla, Goguryeo and Baekje). There were also smaller kingdoms and tribal states. These smaller kingdoms included Gaya, Dongye, Okjeo, Buyeo, Usan, Tamna, and others. From various interchanges there became a prominent north and south division as time progressed.

Japan fought a war with Korea and forced them to trade with them. Prior to conflict, Korea was known as a Hermit Kingdom, and refused to trade with anyone. Japan in 1895, then fought China at the Korean Peninsula, in the Sino-Japanese War. The military of Japan killed the Queen of Korea. The people from Korea made friends with the people from Russia. In 1905, Japan won the Russo-Japanese War against Russia. In the year 1910, the emperor of Japan made Korea a colony of Japan.

From the division of North and South Korea in 1945, there has always been extreme tension between the two nations, and resentment of the United States. Prior to 1945, Korea was under control of Japan, since 1910. After 1945, the Japanese had to surrender the territory. Korea was divided into Soviet and U.S. zones of occupation, with the latter becoming the Republic of Korea (South Korea), in August 1948.

After the Korean War, North Korea (NK) was friendly with China and Russia, but never was formally allied with either, and became more isolated over time. At first, NK became more developed than SK, and this lasted into the 80's. In that time, the South, went from one military dictatorship to another. In the 90's, NK went into famine (called The Great Famine), and there was a collapse of communist trading partners in 1991. It was basically due to the slow fall of the Soviet Union, but it was also because of the difference in political ideologies not shared by other communist. The famine was blamed on the policies of Kim Il-Sung. NK then became more nationalistic under Kim Il-Sung (leader from 1972-1994), who sought to become more self-reliant, especially after the fall of his allies. In other words, he switched from socialistic policies to nationalism. His beliefs were a philosophy called "Juche" (self-reliance and strong military). Rather than import food he let his own people starve to death.

In the 90's, South Korea under now more stable leadership, became a more desirable country. Meanwhile, NK continued to erode under a dictatorship and nationalism. The dictatorship family are known for many scandals and oppressions of their people. Many of the disputes are border disputes, while many are just an extreme dislike between North Koreans and South Koreans, who have opposite ideologies. At the border (where there is a heavily military presence by both armies), they have stare-downs between the two military's, and a dispute can break out at anytime. North Korea has threated the South many times throughout the past few decades and there have been many escalations as seen on the news.

North Korea is nuclear capable and has hinted that they would use nukes. However, their missile systems are not modern age, and they may not be able to hit the US, but they claim to have nuclear weapons, and have conducted confirmed nuclear tests. On February 8, 2015: North Korea tests five short-range ballistic missiles from Wonsan. The missiles fly approximately 125 miles northeast into the ocean. May 9, 2015: North Korea successfully launches a ballistic missile from a submarine, which traveled about 150 meters. The threats are real but the question is, how capable and willing are they? They have South Korea on edge and are a serious threat to their southern neighbors. This is yet another WWIII scenario.

Chapter Conclusion

When it is all said and done, there are a lot of countries willing to go to war, and it sets up a World War III scenario. In 2014, a Gallup International poll of 68 countries, found the US to be the greatest threat to peace in the world, voted three times more dangerous to world peace than the next country. It could be the US will play the role of Germany in a World War III scenario. The US has high debt, an unstable currency practicing QE, and other market manipulations. Many of the US wars, and conflicts are viewed to be unlawful internationally, and even by many Americans, who don't like no-vote war declarations to police foreign nations. Any of these disputes can drag the entire world into a global conflict.

Chapter 17

Glass Banking Pyramid

With increased debt, the interest payments also increase. In 1877, the Ottoman Empire defaulted after the interest on their debt. 50% of its tax revenue went to fund the interest on debts owed. On the eve of the French Revolution in 1788, the French paid 62% of their taxes to fund interest on debt. Many believe the US is too big to fail, but statistically this is the way it works, when a breaking point is reached.

The FED (US Federal Reserve Bank) is leveraged at 78-1. Banks, in the 2008 crisis, went bust with such leverage (Lehman was at 30-1 before defaulting). Financial services firm Lehman Brothers filed for Chapter 11 bankruptcy protection on September 15, 2008. The filing remains the largest bankruptcy filing in U.S. history, with Lehman holding over $600 billion in assets. The ECB (European Central Bank) is leveraged at over 26 to 1.

The US is in foreign hands financially and is being manipulated through the bond market and derivatives – it is a bubble set to burst sometime in the future. When it does, it will take everything with it, because of the size of the US economy. China and Japan own a good portion of US debt. Both countries have very serious financial problems; they could collapse, and no longer be able to shuffle the massive multi-trillion debts of the United States.

Japan for instance, could go bust should inflation increase, and they have to raise rates to curtail inflation. They would no longer be able to pay even their own interest on debts. The higher rates will dramatically increase the amount of interest on debt, which they are already struggling to pay. Japan is paying 43% of tax revenue to fund the interest on debt it owes. Japan is a struggling nation and has many economic woes. They are a supporter of the US bond market, which is the backbone of the monetary infrastructure.

Japan's Debt-to-GDP is 226% (roughly 100% higher than Greece when they went into crisis). The Japanese are buying $1 trillion a year in bonds. Their economy is 1/3 the size of the US and it would be like the US buying $3 trillion a year of foreign bonds.

In 2008, when the crisis hit, the bond bubble accumulated $80 trillion in debt. It has since grown to over $100 trillion.

Even if owners of debt don't go bust, there are currency wars due to political conflicts. The US, has made a lot of countries angry, with its dominating foreign policies, global policing, and wars. America doesn't have the same desirable traits as it did in WWII and that doesn't help things one bit. You might even consider the US filling the shoes of Germany in WWII, a country with a very powerful military that is making people angry and lots of debt. It can lead to a very scary scenario and unstable leadership. Germany as everyone knows resorted to Nationalism, forced labor, and prison camps. The US currently already dwarfs any country in prison population and continues to pass more and more laws while privately owned prisons grow in number.

Geopolitical instability is popping up all around the world, in Ukraine, Israel, Mexico, all over in the Middle East with terrorism, extremism and religious fascism in South America, Africa, even Europe (Greece, Spain, Portugal, Italy, and France) and the UK. As poverty and unemployment spreads the situation will only get worse, everyone is connected through global banking. People who are out of work and in poverty are more willing to go to war, leadership becomes unstable. In my opinion, things do not look good; the math verifies my assumptions.

China's Stock Market took down the global markets in August 2015. The global markets got routed, as concerns about the world's No. 2 economy mounted, with the Dow Jones Industrial Average DJIA shedding nearly 1,200 points between Friday and Monday's trading sessions. In only a few trading sessions, trillions in wealth were wiped out. Japan owns more US debt than China. If even one of them goes, it will cause economic panic. If both go down the US is surely financially doomed. There's likely to be geopolitical instability involved with these crashing markets as the tensions that existed before will become more heated. Whenever China becomes unstable it causes volatility across the world. That is true with most major economies, but China is a major funder of US debt, and has power to influence the financial markets. When the support goes, it will be a chain reaction. The glass banking pyramid will collapse and shatter.

The big countries, who hold the most influence on the global markets, starting at the top are the US, Europe (as a whole), and China:

Country or Region*	2014 GDP In Millions of US Dollars	Rank
World	77,301,958	*
European Union*	18,495,349	*
United States	17,418,925	1
China	10,389,380	2
Japan	4,616.335	3
Germany	3,859.547	4
United Kingdom	2,945,146	5
France	2,846,889	6
Brazil	2,353,025	7
Italy	2,147,952	8
India	2,049,501	9
Russia	1,857,461	10
Canada	1,788,717	11
Australia	1,444,189	12
South Korea	1,416,949	13
Spain	1,406,855	14
Mexico	1,282,725	15

Interest on the U.S. National Debt Source = GAO

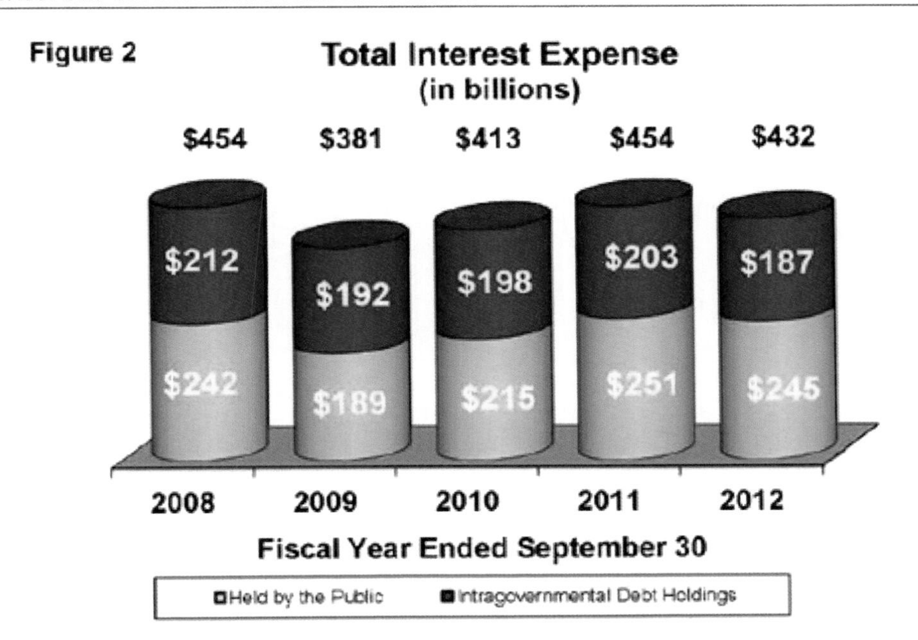

1

Chapter 18

Statistically Speaking

As society grows -- there will be shortages. China and India populations are each 4 times larger than the US. The US is a very large county and uses the highest percentage of resources, because the US has been the richest country for some time. China, however, has increasing wealth, and they are now using more resources than ever before.

The pollution is so bad in some Chinese cities, people wear masks even while in public. All so they could make cheap products for Americans and other rich countries.

China, only has a 3 day supply of coal to power its cities. If they run out of coal there would be black outs. Many Chinese live in extreme poverty. China has hidden debt from factory loans it borrowed to expand their factories and infrastructure. Some economist think China's debt problem is as bad as or even worse than Greece. It is not a pretty situation when you think about it.

US debt is so high, that the national debt combined with the unfunded liabilities, are nearly equal to the total assets of the nation. Some economist believe there is more debt than assets, which means that if the US sold off every corporation, every home, every car, and every asset you can think of, it would not be enough to cover all the debts.

The US, is like a bankrupt corporation that relies on bailouts to function. Much of this is done through manipulation of the bond market by foreigners who control our fate. It causes severe devaluation of American buying power and it causes many people to go into endless debt. If the US were a corporation, an investor would typically look at the debt, and stay clear of it. Since the US receives artificial stimulus, it has not gone under yet.

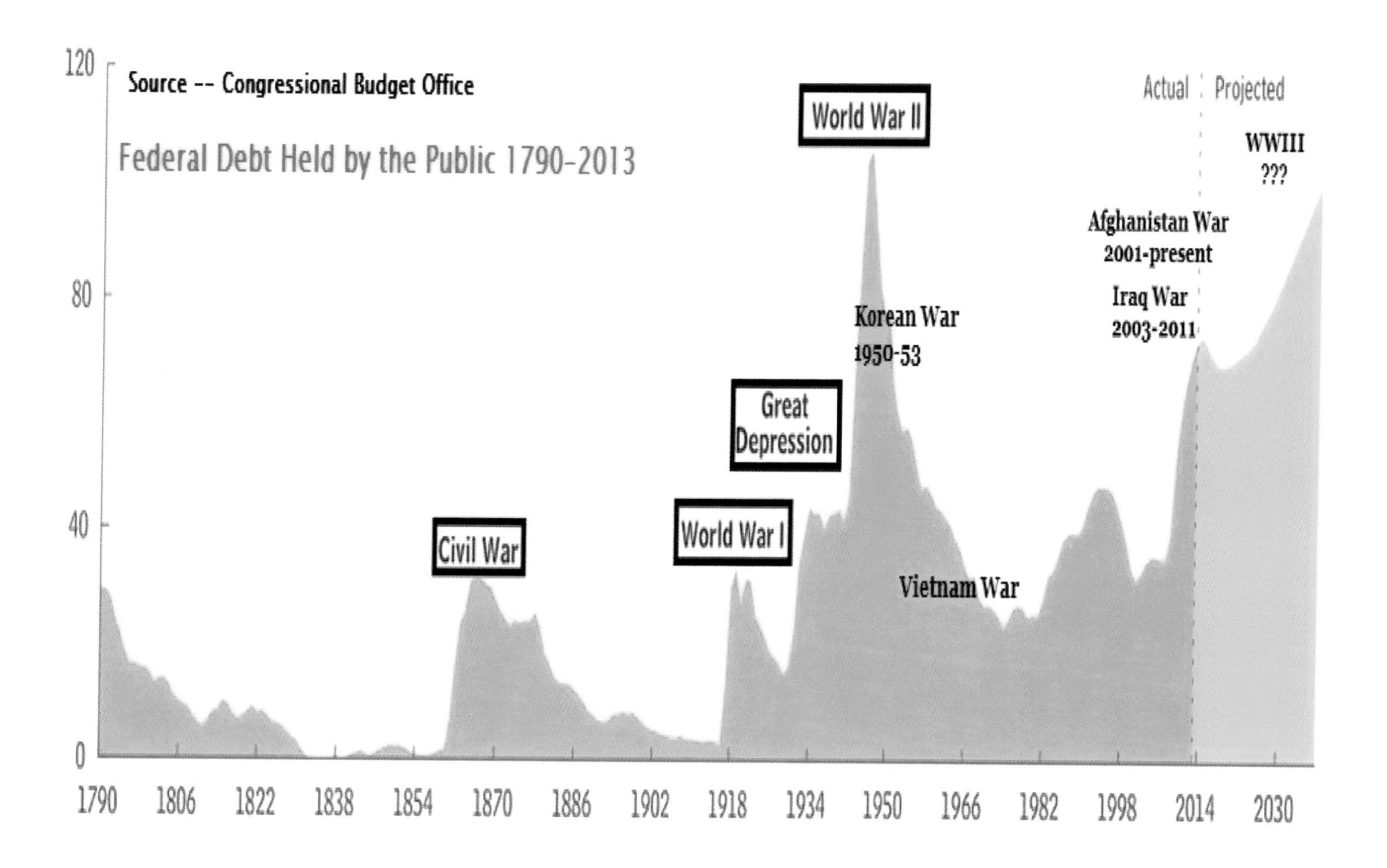
Source -- Congressional Budget Office
Federal Debt Held by the Public 1790-2013
Actual
Projected
World War II
WWIII
???
Afghanistan War
2001-present
Iraq War
2003-2011
Korean War
1950-53
Great
Depression
Civil War
World War I
Vietnam War
120
80
40
0
1790
1806
1822
1838
1854
1870
1886
1902
1918
1934
1950
1966
1982
1998
2014
2030

If you listened to what people say, about how war improved the economy, you might think wars are great. The cost of war is immense and young men die in action in the name of blind patriotism. It causes future inflation due to war debts that we all have to pay for and the weapons destroyed in war are pure inflation. I'm not saying that no wars are justifiable. War declaration, should be a decision not based on might, but on the future of our own existence. Many of these wars could have been avoided before they came to the point of military force.

Even World War II could have been prevented, by occupying Germany after WWI, until they could get back on their feet, and helping them elect stable leadership. Hitler could have also been assassinated – for the price of WWII, I'm absolutely sure it could have been accomplished. Instead it was handled like two childish kids in a playground fight. No side really came out ahead in these wars, despite what is reported by propaganda. The technology gains for people, on the most part, would have been made without war. Technology we have now is spread across the world without wars. People desire things and that is what drives inventions.

Empires don't last forever. Rome had a similar fate, it was engaged in war after war, and created massive debts. Does it remind you of some modern empire? Could it be the United States of America? Even in the days of coinage, they devalued money, by reducing the amount of gold or silver in the coins. The Roman coinage became worthless as a result, because it was the gold or silver that gave the coins value. Roman conquest was so spread out at one point; they could not pay its soldiers because of the cost of all the occupations. In my opinion, that is what collapsed Rome.

What goes straight up must come down to sustainable reality. The US is following in the footsteps of Rome. Most countries have a presence of US military, in the form of bases, or occupations. The US considers countries hostile that do not agree with US foreign policies and they tend to get provoked. It is an unhealthy recipe that may end in disaster, if the pattern of debt and global military expansion continues. I personally think it is too late.

At this point you might think I'm anti-American, but I'm just looking at the problem realistically and statistically. The numbers are bad so the probabilities paint a story. There are two theories I follow and I combine them into one. They are confluence (of factors) and convergence (of trends). When there's a confluence of factors the trends converge together, like rivers meeting together -- the odds paint a certain outcome. When the rivers all meet the course is certain. These patterns include: financial patterns (unemployment, debt), increasing geopolitical conflicts, many wars (war debts, over engagement, new crisis emerging from them), the greater probability of war as the conflicts evolve, and the rise of bad leadership as seen throughout the world. When things are bad – the bad tend to rise. It brings out the worst of human nature, when options are limited; people become more primitive and desperate.

Chapter 19

The Consequences

The likely consequences are: World War III and a massive dark age – The New Dark Ages. I envisioned this while researching throughout the years. Using mathematical theories I had developed to make money, applying the past and future together to plot outcomes. That's what I came up with.

Financial institutions will collapse and wars will break out. I believe these things are already in the making and I thought I might not even get my book published before it happened. It is hard to say which will come first; both could happen in tandem, a financial event trigging a war or a war that collapses society.

The global depression in 1929 preceded World War II, but it took some time. Things seem to happen faster in the modern ages, especially with new technologies, and growing populations. The US may also try to hide the economic collapse with a war, to make it seem like that is the cause of the financial collapse, rather than their incompetence.

Electromagnetic propulsion (EMP) weapons are secrete weapons that are possessed by countries such as China, Russia, and the US. They can shut down all power in a country and destroy electronics. In a World War these weapons will probably be used. It will cause massive chaos and disruptions in how society functions – if it will function at all. You will have no power what so ever, because the electronics will be fried, and the grid will be destroyed. They will not be able to replace all the transformers and so on.

An EMP attack could send a major city back to the Stone Age, or knock out an important military installation instantly. According to National Geographic Channel: "The device would explode at high altitude above the U.S. and release a burst of radiation that would interact with the Earth's magnetic field and atmosphere—including the ionosphere, the thin upper layer filled with free electrons, which facilitates radio communications. As a result, a powerful electrical current would radiate down to the Earth and create additional currents that would course through man-made electrical circuits as well."

The problems are both economical and geopolitical. It stems from greed and human nature, tribally fighting over resources and a difference in beliefs. The technology is misused and the so are the resources. Even though there could be a fair division of resources, artificial scarcity is created by large, greedy corporations. Most of all it stems from the misuse of money, the majority of which is held by a very small percentage.

The next World War will be much different than any other. It will not be a simple trench war like in the past, or air war as seen in WWI and WWII, things will get very dirty and very ugly. Militaries will do things such as shut down the internet, the electric grid, and use very dangerous weapons that many countries now possess. If it goes Nuclear, it will obviously be devastating, and perhaps be the destruction of life on Earth as we know it. Many people believe we are an invincible species and cannot be wiped out. In terms of the Universe, Earth has a rather small imprint in space. There are over 6 septillion estimated stars in the Universe and many believe we are not alone. Earth is a grain of sand in space and we do possess the power to quickly destroy our habitat -- life would no longer be possible.

The next collapse will also be different than any other, because of the extreme amount of debt in circulation. The US, who used to be a highly regarded nation around most of the world, has lost its appeal because of bombing every nation in sight that is considered a threat. So many nations around the world both fear and resent the United States.

World War III is a mostly likely probability and the US could find itself in the role that Germany held. Even if the US does not fill the role of German perfectly, many countries are willing to fight the US, and they are not all in the Middle East. Russia, China, South America, and possibly even Mexico could all join forces against the US. If you believe US propaganda, you believe you are indestructible. Germany once had that belief.

Chapter 20

How it should work – Conclusion

Most countries have government's that are too large, there's too much taxing, and they create too much debt. Most governments should be reduced by 50% or even more. Some unstable governments need to be bigger, but that is usually not the case with developed nations. Spending borrowed money causes inflation, overproduction, and often outsourcing of jobs. As the money supply increases, so does inflation or rather buying power is reduced. Every large-scale injection into the economy has been tried and failed repetitively throughout history. In the 1980s and 1990s, the federal government was cut by one-fifth as a percentage of gross domestic product (GDP) and the U.S. economy experienced a great expansion.

Continual war causes continual debt. Production increases for war are lost in the explosions of the products made (bombs, missiles, equipment, ammo, are all net losses) and deaths of our young. All those resources are lost. War should be limited at all cost.

In a free market system, there is competition, and the government does not intervene. That means bad businesses go out of business, they are not bailed out when they mismanage money, and are replaced by more efficient businesses. If you want deregulation that is what has to happen. Otherwise you keep the regulations in place so the system doesn't get out of hand. It can't work both ways and the rules can't be changed on the fly to make up for severe mismanagement. Corporations get tax breaks, while the people of society are taxed, and have to pay for the corporation's welfare. That is not a free market. It is a corporate dictatorship.

The problem exists at the political level. Corporations are basically hiring politicians and controlling congress through lobbying to influence policies for their industries. There is a conflict of interest between the people and corporate greed, this should not be. Corporate greed seems to win 99% of the time. A politician should not be like a CEO of a corporation; they should have no ties and be for the people.

True growth does not occur unless people have money in the bank, so if you want society to grow, allow the people to save. Buying on credit creates booms and bust. The gains are artificial and the larger the boom the harder the fall. Just look at all the foreclosures and bankruptcies after 2008. Poverty becomes widespread when corruption rules. Wealth gets concentrated into the hands of a few, rather than spread out.

An economy cannot truly expand and expect a positive long-term outcome, unless there is an excess of something, such as a resource. Society cannot expand on massive debts and not expect to experience great poverty. In early times, there was more potential for growth, because there was an excess of oil in the US and other resources. Now almost all Americans have a car and energy resources are gobbled down. One gallon of gasoline is said to represent approximately 500 hrs of human labor and crude oil represents roughly 50% of our energy use. If that source is disrupted for any length of time everything stops.

Society is very populated with very condensed population centers. Continual growth does not work and society needs proper management, which includes the proper use of technology, and investments in alternatives to resources that we are exhausting as a society. Even with the alternatives, society has over expanded at a rate that it would take such a rapid expansion of alternatives; they would not be practical at this time. Even nuclear energy would have to be increased by 2000 fold for the US to be independent of foreign oil. Many are not aware there is also what is called peak uranium. There is only high grade uranium available in limited quantities in Canada; the rest is the bottom of the barrel. Coal may be a good idea for an energy source because the US has ample supplies, but the US has basically outlawed it, while at the same time outsourcing labor to China, who uses it in high quantities.

From a tribal perspective, if one society has an excess, and the other society needs the resource; it usually results in a conflict. If we as humans fail to use the resources properly and do not share them fairly; there will always be war. Since humans are very tribal in nature, this aspect will always exist, and we have to deal with that fact, and the fact that there are haves and have not's that require what we have. The solution is: we must evolve or cease to exist. Even though technology has evolved, we have failed to evolve in our hearts.

With the all or nothing gambling attitude of Americans and our leaders we could very well end up with nothing in the future. It may not be the end of the world, but it could be the end of America. With the massive amount of lobbying occurring in Congress, by representatives of multi-national corporations (to influence our laws), it does not set up a positive future. The US should be concerned with society's infrastructure, the proper use of resources, and the desirability of society. America does not produce enough, but creates massive debts, and engages in endless wars.

The ball and chains attitude is driving the country down. The government's bad economic policies uses and forces us to accomplish its poorly chosen objectives -- rather than by intelligent reasoning. As it stands, the US is a dead nation, that will likely fall on the brink of an economic collapse, and a world war that will spur The New Dark Ages.

Credits & Attributes

Some of the information (like definitions) and the pictures was acquired through Creative Commons Attribution-ShareAlike License of Wikipedia, and wikiquote.org for some of the quotes in the Great Depression.

Most of the information is from my own independent research and notes I've taken over the years -- to develop my theories that go back to 2003, when I intensively started studying the financial markets on a daily basis.

Also, from a wealth of notes from my own personal economic blogs, by daily study of the markets and political developments. The financial information comes from government data, unless otherwise stated in the literature.

https://www.wikipedia.org/
https://en.wikiquote.org/wiki/Main_Page

About Me

Why I wrote this book

I wrote this book to warn people about the coming future -- which I seen in visions. Back in 2005, I told friend the economy would collapse -- housing would pop, gold would skyrocket, and there would be a new great depression.

He looked at me as though I was severely ignorant, because at the time the market was in a raging uptrend. I continued to say, that the economy would collapse, and no matter what they did it would not recover for over 20 yrs and maybe even longer. This person once owned a corporation on the NYSE (in real estate) and acted as though I was nuts.

He asked me why I thought all that? I told him from extensive study of The Great Depression and watching society repeat the same mistakes. He asked, "what mistakes" ? I continued on to state that the US has a negative savings rate, American's spend money on credit that they can not pay back, and the bubble will collapse soon. When it does it will be a worse crisis than The Great Depression, because the bubble is worse.

Another prominent person asked me, why I thought gold would skyrocket? I told him, I can tell by reading charts, because that is what I know how to do and I explained the long term break out pattern that had broke through a key resistance, had built support, and was set up for a major surge much higher.

He said, "Well, I've traded gold with my father since I was young and gold will never go above 600 again." I was trying to gain an investor to invest in gold futures and I actually think he was short gold then, so he was mad I said that.

He acted shaky everytime he seen me after that. I met with a lot of wealthy people, but I could not convince them to help me invest. I was not charismatic enough, but they all lent an ear because of my knowledge. As I followed the market everything I said came true and I had dreams that revealed the future fate of humanity. That is why I wrote the book.

Additional Comments -- Notes

Conversations I wrote in chat on Facebook not added to the book. The content is a bit of a mess but I thought I would include it because it is useful information.

October 24, 1929 - The Stock Market crash of 1929 begins which leads to the Great Depression of the 1930s. It takes 25 years for the Dow to regain its September 1929 high of 381 points.

The unemployment in the US is greater than it was during the great depression. The US continues to stay at war... has runaway debts. but continues to engage in conflicts. The USD is now unsupported (gold confiscated – policy changes) -- the only thing supporting it is a fraudulent global market run by foreign enemies of this country.

Australia

Your debt / stability is tied in with the US and the rest of the world.. I've studied most markets / Australia is mining based and basically walks hand in hand with China. New Zealand is dairy farm land and a financial district. While both countries down under are in far better shape than the US, most countries in the world are still tied into the interconnected global banking structure --- a pyramid of glass set to shatter.

When one major country falls they are all going to go at once and everything will be different. There will be no stopping it. The more it is propped the larger the reset collapse will be. If something doesn't happen you'll turn in to North Korea like dictatorship because that is what happens when nobody steps in to stabilize the condition before it gets out of hand. The US thinks it is invulnerable, as do many major countries that have destroyed themselves by making stupid financial deals.

A little more about me

I've researched the financial markets for 20 yrs and I predicted the 2008 crisis. It was partially a vision I had and through research of human nature and past financial events. I also used a lot of charts and applied my own theories which are something I keep in detail to myself. I'm an expert at reading charts. I have my own system and I used to sell financial software on eBay but there wasn't much market for it. Small traders get angry if they don't make an instant million. If anyone could make instant millions they wouldn't sell financial software but that's what they expect. I also wrote some code for a trading company. The program was called Heartbeat to track when the market ticks so the trading company would know if the data was interrupted.

In my prediction in 2005, I predicted the price of gold would skyrocket, financial markets would collapse / especially housing, and there would not be a recovery. We would go into a deep depression because the crash was around housing (a big ticket necessity) it is far worse than The Great Depression.

People like to argue with me about this but I beat them up pretty good verbally with facts and most don't have comeback arguments to facts. Instead they resort to attacking to make themselves think they are right. They think they win the argument because they get some short-term cycle right that looks like a recovery only to find out it was a lie from fraudulent manipulation but I end up being right in the long run.

The statistics are all basically lies -- fabricated data. Even the way they calculate GDP is a fabrication because they include potential income in the figure and on the most part that potential income is never actually produced but it is still factored in.

John Kennedy (JFK)

What I don't like about him is he tried to change how unemployment was calculated by removing discouraged workers from the calculation. Clinton finalized it by making people who gave up not part of the unemployment figures. If you calculated unemployment the same as in the great depression the unemployment is now greater than the great depression and has been so since Obama took office.

By changing how unemployment is calculated, it paints a different picture on the economy (which is artificial) and makes the statistic useless to past data. The whole point of having data is to compare it with other past cycles to gauge the economy – if it continually changes -- graded on a curve so to speak-- there's no point in having data because there is a new starting point of calculation. Like the debt ceiling you can't just raise the debt limit.

Debt Ceiling (credit card limit)

Just imagine being able to set your own credit card limit. You start at 500 bucks then increase it to 17 trillion without improving your credit worthiness. The government continually raises its credit limit because the government cannot pay itself – the US is therefore bankrupt and has been since the creation of The Debt Ceiling.

If the government didn't raise its credit limit people in the government including police would not get paid. It is a big scam in my opinion because we as taxpaying civilians have to play by strict rules the government itself is immune to. You are basically being treated as a child by the government while the government behaves incompetently.

QE

QE has never worked.. It is just manipulation to prevent an all out crash and a last ditch attempt to help an economy. Japan is still screwed from their housing crash and they have flat rates. They have never truly recovered from the 80's. If they have to raise the rate to stop inflation they will default do to the interest on debt and that is where they are stuck – at the bottom of the ocean on life support. Many countries are in bad shape. China is also near catastrophe. They both fund the USD though bond purchases -- that will collapse.

Albert Einstein — 'I know not with what weapons World War III will be fought, but World War IV will be fought with sticks and stones.'

Thank you for reading! Please rate and recommend. Be sure to share with friends.

I will probably make a remake near the end of 2016.
Sincerely, Neal Vanderstelt 2015

About The Author

I had visions of the future back in 2005. I believed the housing market would collapse (it did!), gold would skyrocket (it did!), and there would be a New Dark Ages (hello! Here we are!).

As a trader of various financial markets, I decided to blog my market sentiments. As I learned more about the money world, I found out how deceptive it all is, going back to its original roots. As time progressed I became an expert at forecasting what would happen next. My methods included reading charts and analyzing economic data. If everyone knew what goes on, they would probably take all their money out of the bank.

I currently trade professionally.

Printed in Great Britain
by Amazon